Securing the Future

Future

Strategies for Exponential Growth Using the Theory of Constraints

The St. Lucie Press/APICS Series on Constraints Management

Series Advisors

Dr. James F. Cox, III
University of Georgia
Athens, Georgia

Thomas B. McMullen, Jr.
McMullen Associates
Weston, Massachusetts

Titles in the Series

Introduction to the Theory of Constraints (TOC) Management System
by Thomas B. McMullen, Jr.

Securing the Future: Strategies for Exponential Growth Using the Theory of Constraints
by Gerald I. Kendall

Project Management in the Fast Lane: Applying the Theory of Constraints
by Robert C. Newbold

The Constraints Management Handbook
by James F. Cox, III and Michael S. Spencer

Thinking for a Change: Putting the TOC Thinking Processes to Use
by Lisa J. Scheinkopf

Securing the Future

Strategies for Exponential Growth Using the Theory of Constraints

Gerald I. Kendall

The St. Lucie Press/APICS Series on Constraints Management

APICS®

The Educational Society for Resource Management

SᵗL

St. Lucie Press

Boca Raton   London   New York   Washington, D.C.

Library of Congress Cataloging-in-Publication Data

Kendall, Gerald I.,
    Securing the future: strategies for exponential growth using the theory of constraints / Gerald I. Kendall.
        p.   cm.
    (St. Lucie Press/APICS series in constraints management)
    Includes bibliographical references and index.
        ISBN 1-57444-197-3
        1. Business managment—Theory of Constraints. I. Title. I. Series.
    RG103.E6R62 1997
    872.5′.26123—dc21   97-61012
                                                                              CIP

© 1998 by CRC Press LLC
St. Lucie Press is an imprint of CRC Press LLC

No claim to original U.S. Government works
International Standard Book Number 1-57444-197-3
Library of Congress Card Number 97-61012
Printed in the United States of America   2 3 4 5 6 7 8 9 0
Printed on acid-free paper

Contents

v

PART III: APPLYING THE THEORY OF CONSTRAINTS TO SECURING THE FUTURE

PART IV: CASE STUDIES

Acknowledgments

I am deeply grateful to my wife and partner, Jackie, whose collaboration and support made this book possible. Faced with the decision of whether or not to continue my effort, I always found Jackie ready to sacrifice and help so that I could complete this work.

Drew Gierman, my publisher, has been an ongoing source of wisdom and encouragement. His input helped a great deal in clarifying the concepts and making the book more appealing to a general audience.

I owe a world of thanks to Bill Dettmer, who referred me to St. Lucie Press, provided invaluable feedback on the book and on Theory of Constraints concepts, and who is an encyclopedia of knowledge on all aspects of TOC.

Many people willingly collaborated on my research and case studies. I am especially indebted to Kim Allen and Wendy Howse from Scarborough Public Utilities, Yvon D'Anjou from Alcan, Jeff Grubb from Orman Grubb, Sam Pratt from Rockland Manufacturing and Daniel Hamilton, who all gave willingly of their time and energy to help make this project a success.

Peter Cassidy from Nortel kept me up to my eyeballs in reading on all of the latest management techniques, and generously shared his analysis of their complementary relationship to the Theory of Constraints.

Peter Urs Bender from the Achievement Group gave me invaluable advice on how to complete my book. As one author to another, over a breakfast meeting last year, he patted me on the shoulder and said, "Gerry, just take a week off, don't answer any phones, don't sleep and get it done!" Well, Peter, I have to tell you that it took a lot longer than a week; I didn't sleep! I got it done with many thanks to you for helping me realize the amount of dedication and concentration required.

There were several individuals who taught me a great deal over these past few years of working with the Theory of Constraints. Dr. Eli Goldratt is not just a genius, but is also a great communicator. Oded Cohen, a deep thinker and philosopher, has continually challenged my thoughts and conclusions. He is a brilliant role model to anyone who wants to believe that the Theory of Constraints is useful to improve everyday life. I commend the APICS Constraint Management Group for their help in spreading the knowledge and practice of the Theory of Constraints. I especially appreciate their recognition of the importance of strategic issues at the executive level.

Finally, I wish to thank my 81-year-old mother, Bea Kendall. A while back she said to me, "Hey Gerald, they've invented a new word for 'downsizing'. They're calling it 'restructuring.'" She never lets me forget that many good people in organizations behave badly, and we can do much better. Perhaps being engendered with this quality of continuously trying to improve is a mixed blessing. I also remember proudly bringing home a math exam as a child, with a 99% on it. Mom asked,

"So how come you didn't get 100?" Thanks, Mom.

About the Author

Gerald I. Kendall is President and founder of MarketKey Inc. (U.S.) and MarketKey Inc. (Canada), international firms specializing in organizational improvement using the Theory of Constraints methodology. He is also a facilitator with York University's Schulich School of Business. Since 1968, he has worked as an employee, manager and executive with small companies and Fortune 500 organizations. He has also founded and grown several companies.

As a former associate of the Goldratt Institute, Gerry completed over 60 days of formal training on the Theory of Constraints and related applications. He is a "Jonah" (trained in applying the Thinking Processes) and "Jonah's Jonah" (expert at teaching the Theory of Constraints), and was trained as a facilitator for Theory of Constraints workshops and application workshops in production, distribution, setting the direction of the company and management skills.

Using the Theory of Constraints as a methodology to address issues of major change, Gerry works with a wide variety of industries. Recent examples include Nortel (Telecommunications), MDS Health Care, Scarborough Public Utilities, Hendrickson (Spring Manufacturing), Dofasco Steel, Alcan Smelters and Best Foods (Manufacturing and Distribution).

About APICS

APICS, The Educational Society for Resource Management, is an international, not-for-profit organization offering a full range of programs and materials focusing on individual and organizational education, standards of excellence, and integrated resource management topics. These resources, developed under the direction of integrated resource management experts, are available at local, regional, and national levels. Since 1957, hundreds of thousands of professionals have relied on APICS as a source for educational products and services.

- **APICS Certification Programs** — APICS offers two internationally recognized certification programs, Certified in Production and Inventory Management (CPIM) and Certified in Integrated Resource Management (CIRM), known around the world as standards of professional competence in business and manufacturing.
- *APICS Educational Materials Catalog* — This catalog contains books, courseware, proceedings, reprints, training materials, and videos developed by industry experts and available to members at a discount.
- *APICS — The Performance Advantage* — This monthly, four-color magazine addresses the educational and resource management needs of manufacturing professionals.
- *APICS Business Outlook Index* — Designed to take economic analysis a step beyond current surveys, the index is a monthly manufacturing-based survey report based on confidential production, sales, and inventory data from APICS-related companies.
- **Chapters** — APICS' more than 270 chapters provide leadership, learning, and networking opportunities at the local level.

- **Educational Opportunities** — Held around the country, APICS' International Conference and Exhibition, workshops, and symposia offer you numerous opportunities to learn from your peers and management experts.
- **Employment Referral Program** — A cost-effective way to reach a targeted network of resource management professionals, this program pairs qualified job candidates with interested companies.
- **SIGs** — These member groups develop specialized educational programs and resources for seven specific industry and interest areas.
- **Web Site** — The APICS web site at http://www.apics.org enables you to explore the wide range of information available on APICS' membership, certification, and educational offerings.
- **Member Services** — Members enjoy a dedicated inquiry service, insurance, a retirement plan, and more.

For more information on APICS programs, services, or membership, call APICS Customer Service at (800) 444-2742 or (703) 237-8344 or visit http://www.apics.org on the World Wide Web.

An Opening Word

*S*ecuring the Future is a survival guide for organizations and executives. It's about the strategic and marketing issues an organization must address if they intend to be around for the long term. It answers the question of how to create exponential growth, by simultaneously exploiting several avenues of continuous improvement. You will discover the measurements that drive a secure, prosperous organization and how to make the transition in a foolproof way. Using a process called The Theory of Constraints, I show you how to dramatically increase the probability of a secure future for the long term.

When you think about ways to dramatically improve, consider these figures. They show the realistic potential of bottom line improvement from different strategies:*

- Cost cutting   1–10%
- Productivity   5–20%
- Automation   10–50%
- Throughput (Marketing)   50–5000%

From 1992 through mid 1996, 163 CEO's of the Fortune 500 must have been working on the wrong strategies, because they were fired!** Most CEO's during that period were focused on cost-cutting and re-engineering (the first two lines in the above list). Why do these strategies often fail in the long term? *Securing the Future* answers this question and offers a systematic approach to really improve your organization — the Theory of Constraints.

---

* Mac Ross, Ross & Company, from a speech at a public seminar, February 2, 1997.
** *USA Today,* April 22, 1997.

1

Part I of *Securing the Future* provides a comprehensive overview of the Theory of Constraints (or TOC for short). It illustrates the step-by-step approach to improving an organization. It's five logical thinking processes move us from the world of opinion and emotion to the world of common sense. In this TOC world, everyone in different functions and at all levels in an organization reach consensus on what makes sense.

You will find many real-life examples of each of the Five Thinking Processes throughout the text. Briefly, the Thinking Processes are:

- **Current Reality Tree** — A process to separate symptoms from their underlying causes and identify a core problem — the focus of our improvement efforts.
- **Conflict Resolution Diagram (Evaporating Cloud)** — A technique that shows why the core problem was never solved, and fosters a new, breakthrough idea.
- **Future Reality Tree** — The strategic solution to our core problem, identifying the minimum projects and ideas necessary to cause improvement.
- **Prerequisite Tree** — The detailed plan of all the obstacles we need to overcome to implement the ideas and projects in our Future Reality Tree
- **Transition Tree** — The actions we need to take, and why, to fulfill our plan.

What is this process called the Theory of Constraints, and why should you be interested in it?

You should be interested in TOC because of the results that organizations are achieving — examples of which are documented in seven case studies in Part IV of this text. Whether you are a multi-billion dollar giant, like Acme Manufacturing, or a $35 million manufacturer like Orman Grubb, or a public utility like Scarborough Public Utilities Commission, you will find issues and solutions in the case studies that you can relate to. I am indebted to the seven companies documented in the case studies which give credence to using the Theory of Constraints.

You should be interested in TOC because many people are focused on the wrong problems, which means they are wasting time and money and falling behind their competitors.

You should be interested because some people in your organization have the right answers but are failing to implement them.

Theory of Constraints (TOC) companies have successfully made several paradigm shifts. These paradigm shifts mean that their organizations are culturally different from non-TOC organizations. A model that helped me understand this when I was first introduced to the Theory of Constraints is the following. It describes two organizational cultures, represented by the two curves of change below:

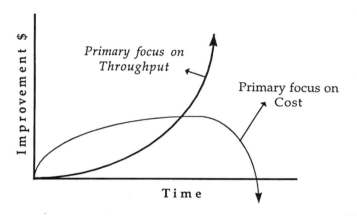

Figure 1   Throughput Model

Throughput has a specific definition within the Theory of Constraints. It means Total sales revenues minus directly variable cost of sales (usually raw materials).

To understand *primary focus,* look at the amount of time spent in meetings and out of meetings, focused on either issues of cost or Throughput. A primary focus on Throughput includes dealing with issues of customer satisfaction and value, and how to increase Throughput through employee productivity and creativity. Primary focus is directly related to measurements focused either on reducing cost or building Throughput. Our primary focus is also determined by how company policies drive behavior and where we invest our training dollars.

On one curve (where the primary focus is on cost), there is quick, dramatic improvement, typical with huge cost-cutting efforts. This is eventually followed by stagnation and worse. Witness organizations such as Apple and AT&T that go through this exercise repeatedly. In this culture, executives never find the deeper understanding of why problems perpetuate, nor what to do about them. Rather, the focus is to get the results needed this quarter, this month, this week to the exclusion of building towards a long-term, secure future.

On the other curve (with a primary focus on Throughput), I typically see a slower start to improvement. To increase Throughput using the Theory of Constraints, you spend time in understanding the underlying problem(s) that is blocking you from improving. This slower start reflects more time spent on analysis, figuring out the complete solution, doing research with customers, gaining buy-in and participation from employees and suppliers, launching and checking results. This is quickly followed by exponential improvement.

It's not at all obvious, as you will see from Chapter 2, which world you are in nor why one paradigm works so much better than the other. If it were, your organization and every other one would be spending 10 times the energy on issues of Throughput than on the other issues.

Organizations stay on the Cost Curve because executives face huge conflicts as they try to improve results. Executives are measured on bottom-line results, as well as other factors. There is a continual conflict between short-term and long-term goals. Yes, we want to have a long-term competitive advantage, but don't invest any money this month — we're reporting our quarterly (or year-end) results.

Another reason is that organizations have three major ongoing sources of disease, that actively prevent them from building Throughput. These three diseases — bad measurements, profit-destroying policies, and lack of correct training also are responsible for terrible organizational conflicts. Why and how this happens, and how to prevent it are topics introduced in Chapter 5, "Three Diseases Attacking Every Organization". Solutions are described in Part III, "Applying TOC" and illustrated in Part IV, "Case Studies."

Competition puts enormous pressure on many organizations to remain on the Cost curve.

For example, Jack Welch, Chairman of G.E., says, "If you can't sell a top-quality product at the world's lowest price, you're going to be out of the game."* Being more competitive often forces you to produce more products, improve them more often and maintain much shorter product cycles. Product development and customer service dollars end up being spread more thinly. This often drives customer satisfaction lower. Many companies put out new products that are more complex and defect-ridden than their predecessors. Part III of *Securing the Future,* and in particular Chapters 13 and

---

* *Fortune,* January 25, 1993, from Jack Welch's "Lessons for Success."

14 on the Supply Chain, suggest a way to view customer needs and competition differently.

To get on the Throughput curve, you must look for better ways to satisfy customers and provide meaningful value. One huge problem today is that customers seem impossible to satisfy. The American customer satisfaction index tumbled down 2% in 1996.* Traditional techniques such as focus groups and surveys aren't telling companies what's wrong. Only a handful of the 200 companies surveyed rated substantially higher with customers than in 1995.

One factor that has driven down customer satisfaction is the Cost-curve mentality and the waves of downsizing, with their resulting impact on consumers. How many hours of frustration are spent every month holding on toll-free numbers, waiting for a human voice to answer.

What customer complaints and demands should you address and how? How do you avoid the negative side effects of constant innovations? The answers are detailed in Chapters 10 to 12.

In order to stay on the Throughput curve, we must also find a way to keep unions and employees satisfied. Otherwise, they may undertake actions that lead to customer dissatisfaction, which ultimately destroys Throughput and profit. Unions justifiably want security for their members. This union pressure to spend more money creates conflicts with executives who feel they must represent the interests of the shareholders. The approach to resolving such a conflict determines which curve you will be on in the future.

In non-union environments, I find employees working 50 to 60 hour weeks, looking over their shoulder constantly to see if the ax is getting any closer to their head. Can these people really focus on doing what's best for the customer? Can they realistically care a whole lot about the shareholder? Chapters 15 and 16 provide approaches to change this paradigm.

This book is about change and how to manage it, using The Theory of Constraints. The methodology is the brainchild of Dr. Eliyahu M. Goldratt, the best-selling author of business management books, including the multi-million seller, *The Goal.* His strategies have been adopted by businessmen and women, teachers, government personnel, churches, studied in universities and practiced all over the world. Dr. Goldratt first introduced the Theory of Constraints in the 1980s, with its application to manufacturing planning and scheduling. He later expanded the Theory to address, systematically and holistically, the problems of complex organizations with many interrelationships and dependencies. A physicist by background, and a man whom many

---

* *Fortune,* February 3, 1997, "StoryTelling, A New Way to Get Close to Your Customer."

describe as a genius and a guru to industry, Dr. Goldratt's gift has been to give average people the tools to think like a genius. *Securing the Future* shows you how to apply these tools to strategic issues.

If you are not familiar with the methodology, don't worry. I do not make the assumption that you are an expert in TOC. Nor do I assume that you must abandon any other methodology you are using. In fact, in Part II, I explain the relationship between TOC and other techniques and show you how to drive better results by combining the necessary tools together. If you never use the methodology, you will still derive great benefit from reading the book and thinking about how to address the important survival issues.

You may have many of the right answers already. The challenge you face is how to get everyone else to buy in to your answers. You wonder why you get so much resistance when the answers are so obvious to you. The ongoing rounds of downsizing and cost-cutting (even if it didn't happen in your company) have taken their toll, and left a legacy of skepticism.

Every day, in the newspapers and on television, there are stories that make workers' hair stand up on end. Read, for example, this letter to the editor from Fred Jones, about how the North American Free Trade Agreement works:

> *"If you're considering purchasing a car or truck from an American auto-mobile company, ask if the vehicle was made in Mexico. If the salesman says yes, tell him to knock 40% off the price. Here's why:*
>
> When the automaker was manufacturing in the United States, he paid his workers about $25 per hour, half their Social Security premiums, three-quarters of their group health insurance and all of their group life insurance. He also paid unemployment insurance and worker's compensation, assisted with pension plans, awarded vacation/sick leave and more. In Mexico, it's quite different. This U.S. auto company pays the Mexican worker 40 to 60 cents per hour and that's it. There's no Social Security insurance or group health insurance. So ask the salesman why is it that when the pickup truck was manufactured in the United States, it cost the buyer $18,000 and now that it's made in Mexico, it's still going to cost $18,000?
>
> This is what NAFTA is all about, folks. Congress passed it and told us how great it's going to be. It sure is for company CEO's and stockholders, but not for you and me. We still have to pay $18,000 if we want the truck.
>
> Oh, when the factory closed, they fired thousands of U.S. workers, too. That's business."

In the same issue of the newspaper, an article described the antics of corporate cost-cutter Albert J. Dunlap. Dunlap has been profiled in major business magazines for his job slashing approach to business improvement. His latest cut of half of the 12,000 employees at Sunbeam Corporation was highlighted in the article describing his $1.8 million home and his attempts to get his neighbors to pay some of the $300,000 armed guard security costs.

Even though most workers, today, are shareholders through pension plan investments, they still separate themselves from those CEO's like Dunlap who gain at the expense of job cuts. Employees, supervisors, managers and executives all have different mind sets. Therefore, you must be able to "sell" the correct answers in different ways to address issues of change. Sometimes, the best and most obvious answers are the hardest ones to sell. A unique strength of the Theory of Constraints is in the tools it provides to communicate your solutions in a powerful way, winning collaboration and commitment to the solution. Therefore, I have selected some case studies specifically to illustrate how communications must change to be successful.

In summary, to have a secure future and to put your organization on the Throughput curve, customer needs, employee needs, and shareholder needs must all be met in a way that is profitable for the business. The competitive threat must be dealt with. And as for your suppliers, you can bully them for a time, but they too will find other customers if their deal is not a win-win. *Securing the Future* provides the guide on how to do it all.

Read *Securing the Future* to challenge your assumptions. Strengthen your plan to get on the Throughput curve and stay there. Use it to augment your efforts and obtain an unbiased look at your strategic plans. Learn how to integrate other approaches, such as Total Quality, into a better overall strategy. Discover how to test different approaches in the market, and find breakthrough solutions.

Both by example and methodology, this text is intended to cause a paradigm shift in your approach. If you are already good, you want to stay good. If not, you must improve to stay competitive. Picture your next competitor as an immigrant with a laptop.* Picture them as having unlimited capital, multilingual, better educated than our kids and very, very aggressive.

Secure your future—before it's too late.

---

* Dennis Waitley, best-selling author of *The Psychology of Winning*.

OVERVIEW
OF THE THEORY
OF CONSTRAINTS

1 Inspect What You Expect — The Goal and the Measurement

I f you understand the extent of failure in most Total Quality approaches, and you believe in the afterlife, then you can probably imagine that Dr. Deming's hair must be standing on end. W. Edwards Deming was the American genius who revitalized Japanese industry through customer-focused quality improvement. Deming taught us that the customer is king. If this is true, then why do so many customers across all measured industries feel less satisfied today than a year ago?*

Deming believed that you cannot improve what you cannot measure. We will integrate a Deming approach with the Theory of Constraints methodology. Before you start building your fortress of security, your must define, "What is the Goal?" Only then can you measure where you stand, relative to the goal.

There are several necessary conditions for a secure future, according to The Theory of Constraints. You may perceive that other conditions are required. I have structured this approach so that you can easily customize the criteria in order to measure where you are.

The criteria that you find in the tables below are derived from several sources, including my own experience with clients using the Theory of Constraints, client research and recommendations, and extensive review of current business literature. They are intended as guidelines, to be customized to your industry and your environment. If some of these criteria are not appropriate to

---

* *Fortune,* February 3, 1997, "Storytelling: A New Way to Get Close to Your Customer."

your environment, check your assumptions with others in your organization and adopt some sensible criteria that you can all agree on. Then measure where you are and set your plan to improve it.

The Goal and the Measurement

In the Theory of Constraints (TOC), I try to not use words like "optimize" or "maximize". Even words like "excellent" are seldom heard. The reason is that one of the underlying assumptions of TOC is that *every* organization has a constraint and can be improved. Until you find an organization that has infinite sales or infinite profits, words like "optimize" are meaningless.

Therefore, in setting your organization's goal and measuring where you are, there is one valid objective — to determine which path you are on. Are you on the road to continuous improvement (the Throughput curve as described previously), or are you taking the path to oblivion (i.e., the Cost curve)?

Where are you now compared to last year? Where do you want to be next year? Use the criteria and measurements below. Set your goal for the future based on where you are now.

The criteria set out below include my observations of the results of successful strategies. You need objective criteria to determine your baseline, from which all improvements stem.

The criteria are set out in four tables with the following headings:

- Financial Security
- Customer Satisfaction and Perception of Value
- Employment Security and Satisfaction
- Marketing and Competitive Advantage

Within each table, I have included a list of criteria and a point evaluation scheme. On the right side, enter your score for each criteria, and total it up by area. The entire survey should take you just a few moments to complete.

For each criterion, if you score positive points, that signifies a desirable result. The more points you score, the more secure your organization is for the future. Some criteria have negative scores. Some negative scores are "awarded" for "accomplishments" such as negative cash flow or layoffs. Other negative points are awarded for failure to survey or not knowing the answers. A lack of knowledge in some key areas is significant in itself, since it indicates a lack of focus on factors that could destroy the organization in the medium term.

Financial Security (Maximum score — 13 points)	
Criteria	*Your Score*
■ **Profitability** (in the case of a "for-profit" organization) +1 point if you are currently profitable (i.e., in the last 12 months). +2 points if you have been profitable for the past 2 years.	
■ **Profitability Trend** +1 point if your profitability trend over the past two years is stable or increasing; −1 point if the trend is decreasing.	
■ **Cash flow** +1 point if it is positive; −1 point if it is zero or negative.	
■ **Sufficient current and projected results to satisfy the shareholders** +1 point if the price per share has increased over 6 months ago; +1 additional point if there are more shareholders today than a year ago; +1 additional point if there are more shares issued today than a year ago and the equivalent price per share is higher than a year ago.	
■ **Enough cash on hand to keep paying necessary overheads and avoid layoffs for some defined period of time, but not so much that we are a takeover target** +1 point if you can pay overheads and employees for a year; +2 points if you can pay for 2 years; −1 point if either you do not have enough cash for a year or if you are a public company and you have more than 2 years of cash or liquid assets on hand.	
■ **Share value/employee** — This financial productivity measure indicates whether decisions regarding the number of employees have had the desirable effect from the perception of the shareholders +1 point if it has increased in the past 3 months.	
■ **Amount of $ invested in Securing the Future of our organization** (e.g., new product development, new markets, R&D, employee productivity, customer satisfaction improvement, adding customer value, etc.) +1 point if we know what this investment is; −1 point if we do not know.	
■ **The extent to which the above investment is working** +1 point if ROI on $ invested in securing the future is >0%; +2 points if ROI >20%, −1 point if ROI <0%, −2 points if ROI is <−20%.	

Customer Satisfaction and Perception of Value (Maximum score — 19 points)	
Criteria	Your Score
■ **# of customers** +1 point if you have more customers than a year ago.	
■ **Average $ purchase per customer** +1 point if the average $ purchased per customer has increased from a year ago.	
■ **Average frequency of purchase** +1 point if customers are buying from you more often than a year ago.	
■ **In-depth understanding of and plan to address the core problems underlying the undesirable effects or complaints of the Market** +1 point if you have a project team documenting actual customer problems, wishes and complaints through in-depth interviews; +2 points if you have an analysis of the market's core problems (Current Reality Tree) relating customer complaints to the root cause of those complaints; +3 points if you have a solution (Future Reality Tree, etc.). See discussion in Chapters 10–12 on this topic.	
■ **Customer perception of value** Pick up to 10 criteria that are key to customer perception of value, according to formal customer surveys and interviews. Survey the employees to determine their thresholds, e.g., in a supermarket, one customer criterion is how long they spend in line. Customer acceptable limit was 1.5 minutes. Employee survey revealed employees thought that 7 minutes was acceptable. +1 point if you have the data for both customers and employees; +1 point for each criteria in which customers & employees agree on appropriate thresholds.	
■ **% of customers lost** +1 point if you measure this figure and have a target; +2 points if you meet or exceed your target; −1 point if you are not measuring.	
■ **Customer Life** +1 point if customers are staying with your firm longer, on average; −1 point if their life is decreasing at the same time that total revenue is decreasing or if you don't measure.	

You do not need to agree 100% with these criteria in order to get value from this text. What you must now acknowledge is the necessity to have criteria, to take a measurement to see where you stand and to gain consensus on what areas your team will target for improvement.

Employment Security and Satisfaction (Maximum score — 11 points)	
Criteria	*Your Score*
■ **Layoffs and downsizing** +1 point if you have had no layoffs or downsizing in your company in the past 5 years; +2 points if your industry has had no layoffs or downsizing; –1 point if you have had layoffs or downsizing in the last 2 years; –2 points if the company was profitable when downsizing.	
■ **Employee perception of security** +1 point if majority of employees perceive that their job is secure for the next year, either in your company or in the market; +2 points if majority of employees perceive that their job is secure for the next 2 years; –1 point if you have failed to survey employees.	
■ **Employee satisfaction — earnings** +1 point if majority of employees are very satisfied with current and prospective earnings capacity in your company; –1 point if you have failed to survey.	
■ **Employee satisfaction — control over job** +1 point if majority of employees are very satisfied with the level of control they have over their job-related activities and decisions; +2 points if the majority of customers are very satisfied with your employee empowerment; –1 point if you have failed to survey employees or customers; –2 points if you have failed to survey both.	
■ **Employee satisfaction — local vs. global optima** +1 point if the majority of employees feel that the goals of the company come before the goals of their department; +2 points if the measurements of the majority of the employees include at least one global optimum; +3 points if the majority of employees report no feeling that their measurements are in conflict with each other; –1 point if you have not surveyed.	
■ **Intellectual capital of employees** +1 point if this is measured and reported internally; –1 point if it is not measured, or not reported or not defined.	

Do not worry about absolute targets. For example, some companies spend a fortune and huge amounts of time refining their forecasting systems to get more accuracy. You hear statements like "We will increase our market share by 3 points over the next 12 months and by 10 points over the next 5 years." Expressed in this way, these kinds of statements are nonsensical. I have yet to encounter a company that can control world events to such an extent that they can make such a prediction meaningfully.

Marketing and Competitive Advantage
(Maximum score — 25 points)

Criteria	Your Score
■ **Killer Factor** +1 point if you have identified a factor in which your superiority to the competition is killing them; +2 points if you have absolute proof that the factor has caused you to win business in at least 25% of competitive situations; +3 points if it will take your competition at least a year to duplicate the factor; +4 points if it will take them at least two years to duplicate the factor. Double the points if you have more than one factor.	
■ **Sales Force commitment and stability** +1 point if sales-force turnover is less than 20%; –1 point if you have not surveyed; +1 point if the sales force, on average, is earning more money this year than last year.	
■ **Marketing strength** +1 point if your sales revenue per marketing dollar spent has increased in the past year; –1 point if you haven't measured it or if you do not have a person responsible for marketing.	
■ **Test Marketing** +1 point for each market test you conduct per year, where you formally change your marketing proposition to your clients, up to +5 points.	
■ **Products and services** +1 point if your customers can buy more products or services from you this year than last year, i.e., if you have introduced new products to the market, +2 points if the new products/services are being purchased by more than 10% of your existing clients.	
■ **Market Share in target segment** +1 point if increasing over last year.	
■ **Cost of goods compared to competition** +1 point if you can buy the majority of goods that you resell or your raw materials at the same price as the competition; +2 points if your price is substantially lower for most goods; –1 point if your price is higher than competitors, but within 5%; –2 points if your price is >5% higher than competition for most goods.	
■ **Market Coverage** +1 point if you are 100% sure that you are reaching >80% of your target market with your message.	
■ **New Products and services revenues** +1 point if income from new products and services consistently exceeds or meets targets.	

Marketing and Competitive Advantage (Continued) (Maximum score — 25 points)	
Criteria	*Your Score*
■ **Diversity** — This criterion looks at your target markets and the extent to which your management team has the diverse background to represent it, understand it and address its needs. Looking at the ethnic, sex and other compositions of your target customer base (e.g., black, white, male, female, Indian, Chinese, European, etc.) what % of your target customer base is represented by identical ethnic background in your management team? +1 point if >10% are represented; +2 points if >25% are represented; +3 points if the management team can collectively speak >5 languages.	

Continuing with the example of improving market share, how would the Theory of Constraints approach improvement? First, you confirm that any increase in market share signifies a desirable movement closer to your goal of increased profits. Then, you put a plan in place to make such improvements, in small, continuous steps, without worrying so much about whether the exact figure is 2% or 5%, for now.

Your plan is in the form of a Future Reality Tree (or other strategic plan identifying the projects and ideas necessary to cause the increased market share and other desirable results). These ideas reflect the minimum effort necessary to get at least X% growth. In addition, it includes a contingency plan in case you end up with too much market share or too little. The negatives of too much market share might imply dealing with all of the problems of becoming a monopoly when you don't want to be a monopoly. Or it might mean dealing with the nightmare of companies like America Online, who promise the world and suddenly can't deliver because of a shortage of capacity.

In the Theory of Constraints, all of the potential negative side effects that could result from change are identified *a priori*. Contingency plans, with the input of all affected functions and departments, are laid out with enough preparation so that they can be implemented without a disaster occurring.

In coming up with the strategies to secure the future, many clients wonder how to distinguish a goal from a necessary condition. The answer is that from the point of view of constructing a solution (e.g., a Future Reality Tree), there is no difference. For example, you might say that your goal, in securing the future, is to have an increase in price of your stock greater than the average in your industry. A necessary condition might be having a "killer factor", like the quality advantage that the Japanese used in the automobile industry, that will give heartburn to the competition. Another necessary condition might be that customers buy twice as much quantity as they are currently buying. You could make any one of these your goal and the others the necessary conditions. It doesn't matter. The solution (Future Reality Tree) would look identical. The projects and ideas required to cause the goal and the necessary conditions to occur are identical.

For those of you who are not very familiar with the Theory of Constraints, you may worry about the use of some unfamiliar terminology. I want to make sure that terminology is not an obstacle to your getting value from this book. Therefore, in the next chapter, I provide a brief introduction to the Theory of Constraints, and tell you how *Securing the Future* fits in. In addition, you will find a glossary at the end of this book providing a handy reference to unfamiliar terminology.

2 Two Paradigms — Moving from the Cost World to the Throughput World

There are two unique and very separate worlds that organizations live in — the *Cost* world and the *Throughput* world. These worlds are so far apart that it often takes years to travel from one to the other. The cultures, the measurements and the thinking are so different that you cannot have both worlds co-existing in an organization. Once you make the commitment to enter the Throughput world, you leave the Cost world behind.

Throughput (T) is defined as the difference between Sales and Direct (or Truly Variable) costs. It is similar to Gross Margin, but not necessarily identical. Often, when companies produce and/or sell multiple products from common processes, cost accountants try to allocate common costs to determine product margin. This can grossly distort the true out-of-pocket cost of producing a product and lead to bad management decisions, as you will see from the example below.

When Dr. Goldratt saw executives making terrible management decisions based on how costs were allocated to different products, services, facilities and other entities, he came up with the Throughput concept as a decision-making tool. If a customer buys one more product from me, how many dollars will I deposit in my bank account, assuming that the only people I pay expenses to are my materials and parts suppliers? How much would I really have to spend to produce one additional product?

If we already have a manufacturing plant, tooling and labor in place, our only additional direct costs are the raw material costs. If there are other truly variable costs (e.g., purchased parts or major electricity that we can calculate), then we will add those costs to the direct costs.

Throughput (T) = Sales – Truly Variable (Raw Material & Parts) Costs

All of the indirect costs associated with selling the product, getting the product out the door and collecting the payment from the customer are called Operating Expenses (OE). These include:

- Plant labor
- Administration and other operations labor and overheads
- Utilities
- Corporate overhead
- Depreciation
- Sales and Marketing expenses

The third element of the Throughput world is the investment (I) you make to satisfy your customers. In many companies, the investment is both in terms of the cost of the inventory you carry at all levels and the large asset investment in buildings, equipment, fixtures, etc.

In the Cost world, people focus primarily on decreasing the operating expenses (OE) and investment (I), and less so on increasing Throughput (T). It's not obvious that management is doing so. However, there are many symptoms of which world an organization operates in, and we'll review those shortly.

People who believe they are operating in the Throughput world are often, in fact, living in the Cost world. You just don't realize it. You believe that you are doing the best you can possibly do to increase company profits and, eventually sales. However, when you examine how decisions are made, the criteria are cost-world criteria.

Cost-world criteria are harmful in two ways:

1. **They distract from meeting the goal of the organization** — Given that a for-profit organization is created to make money, the shareholders expect a competitive return on the investment they put into an organization. If they don't get a satisfactory return, they take their money elsewhere.

The kind of return on investment that keeps shareholders around for the long term is not created by cost-cutting. I'm sure that every shareholder appreciates a cost-conscious company. Cost cutting may work short term, but it does nothing to create a solid, long-term high growth stock. If anything, persistent cost-cutting is probably a symptom that management doesn't know what it is doing.

Super happy shareholders are created by growing Throughput and profit, outsmarting competitors and managing growth to keep the customer base happy.

This is not to say that costs should be ignored. In the long run, however, it's a question of where the constraint is that is blocking the company from improving and whether that constraint is driving priorities.

2. **Many Cost-world decisions actually decrease Throughput and Profits**

Let's consider an example.

A huge computer company (call them X), known worldwide for the quality of their personal computer products and the friendliness of their operating system, began losing market share to their competitors.

The competition's products became easier and easier to use. The price of the PC's running the competition's operating system dropped in half almost every 1 to 2 years, while these PC's became more powerful.

In 1995, sales of Company X began declining. Profit began to tumble and the stock price began plummeting. In the final quarter of 1995, the company began to lose money. The president was negotiating to sell the company to a manufacturer of medium-scale computers. When the deal fell through, that president was fired and a new president was brought in from a completely different industry.

The new president, under severe pressure from the board and shareholders to show results, immediately cut budgets across the board. He demanded a downsizing of over 1,000 jobs. Simultaneously, he wanted increased productivity from those employees who remained. Of course, to cover his behind, he carefully explained to the board and to shareholders how such a downsizing would cause a major loss in the current quarter to cover severance packages. No problem, the board thought. The company will be rolling in profits by the following quarter.

> *In the Cost world, if you don't know what else to do to improve your business by next quarter, downsize. If you're afraid of bad press, call it restructuring.*

As you may know, high-technology companies require huge amounts of money to be spent for extensive research and development. Often, the fruits of these labors are not seen for several quarters, and sometimes for several years into the future.

In the Cost world, when managers and executives are forced to make significant budget and headcount cuts, they typically focus on milking the present and sacrificing the future. It's a natural survival instinct. It's like a family that faces a short-term financial crisis. They continue to pay the mortgage and buy food, but stop the retirement savings, life and health insurance, and maintenance on the car.

In the Throughput world, the attempt is made to identify the constraint. All energy is focused there. As Bill Dettmer, author of *Goldratt's Theory of Constraints* told me, "It's stupid to do anything involving changing capacity without knowing exactly where the current constraint is and where the next constraint will be!"

While downsizing and cost-cutting decisions may make sense in the context of a true financial crisis, they do not make sense without first analyzing what the constraint is. These decisions did not make any sense for Company X, and the employees knew it. Company X was not in a financial crisis. They had several billion dollars in cash reserves. These decisions began to develop some serious negative side effects that haunted the company and will continue to haunt them for years into the future.

Sure enough, in the quarter that this new president took over company X, losses were almost a billion dollars from the restructuring.

The president promised a profit the next quarter, but instead faced another loss of close to $40 million. Stunned and disappointed, he began pointing the finger at everyone around him. Key executives started leaving.

He asked for an analysis of profits and losses by product line. This is one of the standard and most dangerous reports that many executives and product line managers in the Cost World use as a tool to manage their business. Some of the newer products were burdened by cost allocations from the entire corporate overhead structure. It made these products look terrible, even though, in truth, there was relatively little actual out of pocket outlay. The vast majority of the allocated dollars would have been spent whether or not these new product lines existed.

The new president immediately decided, with little consultation, to cut out those product lines. He reasoned that it would improve profits tremendously. He mistakenly believed the cost allocations were actual costs that could be recouped. He neglected the impact such cuts would have on both employee morale and customer confidence.

Result: Small decrease in actual costs; huge decline in revenues.

Similarly, he then asked for a *make* vs. *buy* analysis on all internally manufactured products. This company manufactured many of its own components as well as its operating system. On paper, when the overhead costs were allocated to each of the manufactured components as well as the operating system, and compared to what it would theoretically cost to buy these items from outside, it looked cheaper to buy from outside. It just makes sense, doesn't it?

The fallacy in this thinking is that you cannot lay off a plant. You cannot lay off a lot of manufacturing equipment. You cannot lay off desks and chairs. All you can lay off on short notice are people, and if they are full-time skilled employees, layoffs can cost a lot of money in severance packages and future consulting costs.

The president demanded that the plants start buying components from outside that were "cheaper" than the cost to manufacture. He canceled all development on their existing operating system (the simplicity of the operating system was the main reason why customers bought this machine). He used 25% of the company's remaining cash to buy another company and their operating system.

The next quarter showed a dramatic drop in sales, as customers around the world lost confidence in the viability of this manufacturer. With their abandonment of new product development, Company X began to look less and less attractive to many customers who had loyally supported the company for years. Software developers, who had been developing for the Company X operating system, began moving their products over to competitor's operating systems.

Company X stock dropped from $48 per share to $16 per share in just over a year. The company decided to lay off several thousand more people.

What went wrong? What should the company have done and why wasn't it obvious?

This company was #1 in several niche markets of the personal computer world. They had enormous cash reserves in 1995. There was no immediate need to create a spiral of negative side effects by downsizing. There was certainly an urgent need to focus on increasing Throughput.

If you took over a company whose entire success and lifetime strategy was built on providing tremendous productivity and ease of use advantages to niche markets, what would *your* logical choices be to increase Throughput? What major factor(s) would *you* use to build additional market, using your existing strengths in manufacturing and marketing?

This company chose to go head on against their competitors on price. They failed almost immediately. They invested hundreds of millions of dollars to manufacture their own cheap clones, and failed to attract enough additional market to break even.

Is it obvious only to me that this strategy of copying the competition was doomed to failure from the start? I don't think so. I believe that many people inside company X could have shown the fallacies here before they ever began trying to roll out these cheap clones. What they accomplished was to convince customers, who would have bought their more expensive products at the higher price, to pay less.

The easiest strategy in the world to copy is price cutting. Many companies commit suicide doing it. The advantage of any price cutting is usually so short-lived, I would hesitate to call it a strategy. It is symptomatic of companies that don't know how to think through a competitive marketing strategy. If you want to become Walmart, great! How will you block the competition from meeting or beating your price?

The above scenario is somewhat obvious to people who already understand the fallacies of cost accounting and allocations.

What about companies that are making money, but are still in the *Cost World?* How do we recognize those companies and send up a red flag? Here is an example:

Let's assume that you have three major projects to complete this year.

Project A will generate $450,000 in additional Throughput.
Project B will generate $550,000 in additional Throughput.
Project C will generate $650,000 in additional Throughput.

Operating Expenses for all projects is expected to be $1,200,000. Total Throughput is expected to be $450,000 + $550,000 + $650,000 or $1,650,000.

Given the following people resources available to each project, what is the *maximum* Net Profit you would expect to result, in total, from all projects? Note that anyone who is not occupying all of their time on these projects can work on other projects with similar characteristics and generate similar Throughput for the company.

Table 1

People	Project A	Project B	Project C	Total Hrs. Required	Hrs. Available
Tom	30	5	5	40	50
Dick	10	10	30	50	30
Harry	15	20	20	55	60
Thelma	15	25	10	50	75
Louise	3	10	3	16	40
Mary	3	10	3	16	30

Most people will take the Throughput from each project, add it together, and subtract the total Operating Expense. If you guessed $450,000, you may be living in the Cost World, or at least you have some assumptions that do not reflect reality in today's corporate world.

In the Throughput world, we know that we have one or more constraints — something that blocks us from moving closer to our goals. With projects, typically our constraints are people or some other physical resource.

Every single company that I have done business with in the last ten years has more projects on the go than they have physical capacity to complete on time, on budget and within specifications.

The first step in the Theory of Constraints is to identify the constraint. We have six people available to complete the projects. Given the weekly time requirements expected for each project, and the amount of time that each person has available, we can draw a conclusion about whether or not a resource is a constraint:

Table 2

People	Project A	Project B	Project C	Total Hrs. Required	Hrs. Available	Conclusion — Constraint?
Tom	30	5	5	40	50	No
Dick	10	10	30	50	30	Yes
Harry	15	20	20	55	60	No
Thelma	15	25	10	50	75	No
Louise	3	10	3	16	40	No
Mary	3	10	3	16	30	No
TOTAL	76	80	81			

Based on Table 2, we have one constrained resource — Dick. He has 30 hours available per week, and the demand on his time from all projects is 50 hours.

> **CONCLUSION: Dick is our constraint. We cannot complete all projects with current resources. We must choose which project(s) to complete or get more of the "Dick"-type resource.**

I know that this is not the way that most organizations operate in practice. For one thing, they leave all projects on the books, and then put pressure on already overworked resources to work harder. This compounds the problem even further, by making all projects even more behind schedule.

Which of these projects should Dick be assigned to, in order to try to get the best profit for the company — Project A, B, or C?

The common way to decide would be to look at the investment of labor hours required for each project and the total Profit that each project is projecting:

Table 3

	Project A	Project B	Project C
Throughput	$450,000	$550,000	$650,000
Total Hours Invested	76	80	81
Throughput per hour invested (Throughput/Total Hrs. Invested	$5,921	$6,875	$8,024

From Table 3 it is clear that Project C is the best one to prioritize. A Throughput per hour invested of $8,024 would be received from Project C. Alternately, we could do projects A and B and receive an average Throughput per hour invested of $6,410. Therefore, in every respect, Project C looks like the winner. But is this really the case?

Since we often have different functions in the company fighting over the same resources, the likely winner will be the executive in charge of Project C. Is this the best decision for the company? If we consume all of Dick's time on Project C, we will generate $650,000 in Throughput, $1,200,000 in operating expenses resulting in a loss of $550,000!

Since we've identified Dick as our constrained resource for all of the company's projects, what happens when we move into the Throughput world? Will we make the same decision?

The second step in the Theory of Constraints, when dealing with a Physical Constraint, is to exploit the constraint. This means that we want to get as much Throughput as possible from the constrained resource. By definition, every additional dollar of Throughput we can get out of our constrained resource is additional Throughput for the company.

Therefore, our analysis of which project(s) to focus on is analyzed from a different perspective. What we must look at is not the total labor time on each project, but just Dick's labor time on each project. We are interested in Throughput per constraint hour, not Throughput per total labor hour. Knowing this, we can decide how to exploit (take advantage of) our constraint.

Table 4

	Project A	Project B	Project C
Throughput	$450,000	$550,000	$650,000
Total Constraint Hours Invested	10	10	30
Throughput per Constraint hour (Throughput/Constraint Hrs. Invested)	$45,000	$55,000	$21,667

From Table 4 it is apparent that project B is the best choice, giving us the highest Throughput per constraint hour. Since Dick has 30 hours available and this project requires 10, and Dick can do other projects of this type, he could do three projects of this type in the time he has available (30 hours). Looking at Table 1, we can see that if we did three projects of type B, we have enough of all other resources available to handle it.

Our total Throughput from three projects of type B would be $1,650,000. Our total profit from three projects of type B would be $450,000. Clearly, this is far superior to our original choice.

Let's suppose you hadn't done this type of analysis. Two executives approach you with two different ideas. Each idea requires an outlay of $10,000. The first executive asks you to invest $10,000 to buy a specialized computer for Tom. He proves to you that Tom, a highly skilled $100,000 per year engineer, will be twice as productive if he had the new computer. The second executive asks you for $10,000 for overtime money for Dick (a $40,000 per year computer analyst), for another 10 hours per week of his time.

- Which executive is likely to get the money?
- Which use of the $10,000 will have the most positive impact on T, I and OE?

Since Tom is not the constraint, making him twice as productive will accomplish absolutely nothing for the company. Since Dick is the constraint, and assuming he is already assigned to Project C, with the additional 10 hours, he could accomplish Project A and generate another $50,000 in profit for the company — a 500% return on investment!

This is just an example. The point is that common sense — what the Throughput world is all about — is not obvious. Here are some other factors that are symptoms of operating in the Cost world:

- **When you ask executives and managers what projects they have on the go to improve the company, they will list several or many projects. Most of the projects describe efforts to improve their own areas. Most of the projects do not indicate what the expected impact will be on T, OE, and I.**

  For example, the Director or VP of Corporate Communications tells you about projects to develop proactive communications (external and internal) internationally. He describes how he is focused on a team approach between employees and others. He is developing an intranet, an internal means of sharing information instantly between managers, employees and executives. The project was initiated without identifying that the speed of sharing information is a company constraint.

  The Health and Safety Director talks about his proactive projects to align the organization with directives from the President to promote non-smoking in the workplace.

  None of these people talk about how much more Throughput will be achieved if the project is completed, or how much Operating Expense or Investment they are proposing to decrease.

- **There is much more focus on activities than results:** "Wow, our whole team worked 18 days straight, 15 hours per day. We had 36 meetings, and developed an entire infrastructure to support the new communications environment."

  By contrast, *all* of the managers in an organization in the Throughput world talk about their projects and work in terms of moving closer to their Throughput goals. For example, see the Acme Manufacturing case study.

- **Management reports are full of allocated and fabricated numbers:** If you are not sure what this means, read the Acme Manufacturing case study and in particular the controller's comments about Standard Cost reporting.

- **There is significant mistrust between functions and levels:** Typically, the executives are blind to the problem or make statements like, "It's always been that way".

- **Employee turnover is above 20% per year and it is equal to or higher than last year:** This may be symptomatic of companies whose employees do not understand or accept the company direction. It is characteristic of employees who do not see a future in their company, meaning that they don't view the executive's decisions as making sense to them. The turnover percent that you need to worry about is specific to your company.

- **There are more per capita customer complaints than a year ago:** Customers who believe they are getting value from their suppliers and are dealing with employees who are helpful do not complain as much. Companies in the Throughput World also perform extensive examinations of potential negative side effects before rolling out new programs, and eliminate the negatives before they do harm.

The first step in changing paradigms from the Cost world to the Throughput world is to recognize your current reality — which world your organization is in. In Chapter 5, I expand further on the symptoms of the Cost world and how to find the three diseases of the Cost world that are attacking every organization. Before you can fully understand the meaning of the term "disease," as it relate to the Theory of Constraints, you need to understand more about the components of the Thinking Processes that make up the Theory of Constraints. In the next chapter, I introduce these components in a real-life, simple example.

3 Overview of the Theory of Constraints — An Example

The purpose of the next two chapters is to provide you with sufficient background to be able to read and understand the Theory of Constraints diagrams — not just the superficial reading, but the thinking behind them. The intention is not to teach you how to construct such diagrams, since that usually takes several days. By understanding them, you will be able to follow all of the examples more comfortably.

There are many techniques available to analyze and resolve problems. I am not asking you to give up any technique that works for you. The point is that neither management consultants nor Japanese auto makers have bigger brains than you do. The only difference between people who come up with breakthrough solutions and people who do not is the way in which we think about a problem and its solution. Therefore, the ability to look at a complex problem in various ways, using different techniques, is often the reason for a breakthrough solution rather than a disastrous one.

The methodology and the examples you will see come from applications of the *Theory of Constraints*. Numerous books, articles, videos and other formats describe it. The Institute of Management Accountants has published a monograph about it.* Thousands of organizations and individuals have learned and applied it successfully.

---

* *The Theory of Constraints and Its Implications for Management Accounting,* Norzen, Smith & Mackey, The North River Press 1995.

As much as possible, I've written generically (i.e., in plain English), with the assumption that you know nothing about the Theory of Constraints methodology.

Every theory has assumptions, and the Theory of Constraints is no exception. Let's examine a few of these assumptions and see if you agree.

One assumption is that every organization has at least one constraint — something that's blocking it from moving closer to its goal. For example, imagine a company whose goal is to make more money. If they have no constraints, then their profits must be infinite or certainly equal to the total profits of the world for all companies. Imagine a hospital whose goal is customer satisfaction. Can you envision a perfect hospital, in real life? Or a perfect marriage?

The Theory of Constraints assumes that diseases exist in organizations and relationships. Using cause–effect analysis, we can find diseases and eliminate them. Similar to medical practice, if we treat symptoms and not the disease, the disease recurs. To cure the disease permanently, we must find the root problem(s) or cause(s) and eradicate it throughout our organization.

For example, in a retail store, did you ever have to wait in line for several minutes at the checkout because the item that you or someone in front of you purchased didn't have a price or bar code? Store managers and cashiers hate when this happens — they continually get complaints about it.

So what does a store manager typically do? They give a boring lecture at the next staff meeting about the problems of unmarked merchandise. They may even track down the culprit in each incident and talk to them. Or a corporate systems person will look at the problem and propose a new, multi-million dollar computer system to ensure all goods going out to the floor are marked.

However, these approaches continually fail to eliminate the problem *because they are dealing with the symptoms and not the root cause.* The above example is a typical problem where the root cause relates to measurements, one of the three great modern organizational diseases. See Chapter 5 for further discussion of this problem.

Let's work through a brief example, from beginning to end, of the application of the Theory of Constraints to a real life problem.

Let's assume that I am the chairman of a company with several divisions and I've just appointed you President of my largest division, at twice the salary and bonus you earned before. The division manufactures food products and sells to retailers and fast food restaurants.

At the end of your first year, sales are stagnant and profits are down. I can't report lower earnings this coming quarter, so I'm giving you three choices:

1. **Cut costs**
2. **Increase Sales**
3. **Resign**

Most North American managers are action prone. This eventually proves fatal. So let's try a more scientific approach. The first step is to scientifically analyze why sales are down. The Current Reality Tree, the first of five TOC Thinking Processes, shows you how.

Diagnosing the Disease — The Current Reality Tree

As a first step, you need to separate the symptoms from the disease. If you have a brain tumor, it's no use giving you medicine for headaches and nausea. If you have stagnant sales and lower profits, it's useless to treat symptoms.

A lasting solution is based on correctly diagnosing the disease. Learning how to understand your current reality requires two skills necessary to diagnose an underlying problem:

■ **Ability to understand common sense**

With some problems, the disease and the cure are "obvious." Everyone agrees. With complex problems — those where several functions or levels in an organization are involved — what's obvious to one person is nonsense to another. The Theory of Constraints provides Eight Rules of Logic to make sense out of any situation. See Glossary — Categories of Legitimate Reservations.

■ **Ability to construct common sense**

Using the Eight Rules, you establish what's really going on in your organization. You separate fact from fiction. Working from a small list of Undesirable Effects (UDE's), you build a comprehensive Current Reality Tree. The diagram often looks like a tree, with several branches. It shows connections (via arrows) between causes and effects. The connections, if the logic is correct, explain that one or more statements (at the bottom of an arrow) cause(s) the statement at

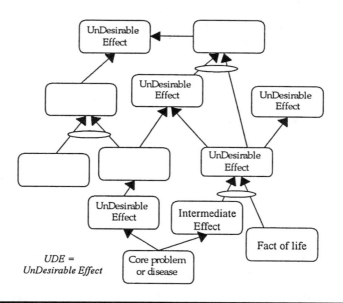

Figure 2   The Logic Tree

the top of the arrow to occur. In a properly constructed tree, you see at least one entry which leads to most of the undesirable effects. This core problem, or disease, is the prime target for the improvement efforts. Get rid of it, and you'll erase all resulting undesirable effects. The logic tree is built in the form shown in Figure 2.

In our real-life example, sales are stagnant and profits are down. What are some other undesirable effects? Here is a list of four:

1. Managers are leaving.
2. Customers are not buying as much as they used to.
3. Cash Flow is decreasing.
4. Demands on customer service are going through the roof.

In our real-life situation, the first step in diagnosing the disease is to understand the symptoms. Then, you make connections, using the Eight Rules of Logic. When you are finished, the diagram reads like a true story. It is connected by IF-THEN logic. You substitute the word "AND" with the ellipse shapes ($\Longleftrightarrow$). For example, you read Figure 3 starting at the bottom of the diagram, as follows:

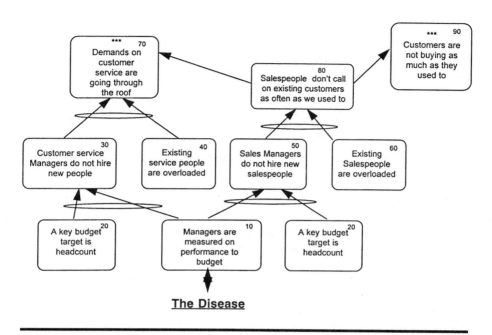

Figure 3   A Current Reality Tree

IF managers are measured on performance to budget, AND a key budget target is headcount, THEN this causes sales managers not to hire new people. This also causes customer service managers not to hire new people.

Reading through a complete Current Reality Tree, you explain how every undesirable effect is caused. Using the rules, everyone can challenge your logic. Anyone reviewing such logic with the "owner" of the tree quickly identifies the assumptions in their current situation.

The remainder of the story (simplified from real life) is read as follows:

> Existing salespeople are overloaded. If sales managers do not hire new salespeople and existing salespeople are overloaded, then salespeople don't call on existing customers as often as we used to. If this is true, then customers are not buying as much as they used to.
>
> Existing service people are overloaded. If customer service managers do not hire new people and existing service people are overloaded, then demands on customer service are going through the roof. But that's not the only reason. Also, if salespeople don't call on existing customers as often as we used to, then demands on customer service are going through the roof.

The Beginning of a Cure — The Conflict Resolution Diagram

In the Current Reality Tree diagram, there is an entity (statement) that connects to both branches of the tree — the customer service branch of negative effects and the sales branch of negative effects. That entity states that "Managers are measured on performance to budget". Is it possible that budgets are bad for your business?

On the surface, the measurement sounds good, not bad. How can it be a disease?

Diseases are hidden in the complexity of organizations — we deal with so many variables, so many motivations and measurements, policies, etc. When we suddenly discover the disease, we often find that it's not something new or unknown to us. We just never connected it before as the cause of so many problems.

Without such a focusing technique as the Current Reality Tree, we diffuse our energy on dealing with many symptoms rather than one core problem. If customers aren't buying enough, let's put on a promotion. If sales are down, let's have a sales contest. This erases the symptoms temporarily, but they always come back to haunt you. Compromise is not a cure.

The Conflict Resolution Diagram, or Evaporating Cloud, in five boxes and a few statements, builds a new insight into the disease. By understanding the underlying conflict preventing a cure, we move to the beginning of a breakthrough — a first injection (idea) to start curing the disease.

The Conflict Resolution Diagram focuses on how to find a starting breakthrough idea. This requires three skills:

1. **Ability to construct a conflict diagram.**
   Defining the problem is half the battle of solving it. The core problem has never been solved before because some conflict has prevented you from either recognizing or attacking it. You only have 5 boxes to fill in, to define the problem. When completed, the diagram is a fantastic tool for communicating to others why the disease has never been cured.

2. **Ability and willingness to bring your hidden assumptions to the surface.**
   Assumptions tell us why we have been behaving in a certain way, or living with a conflict for years. The cloud exposes the assumptions that have been guiding your organization. Some of those assumptions are valid. Others are not and need to be challenged.

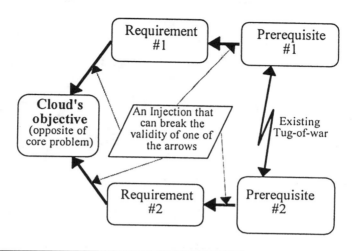

Figure 4   The Conflict Resolution Diagram or Evaporating Cloud

3. **Ability to find breakthrough ideas.**
    Once you know which assumptions are erroneous, you must overcome at least one of them. You are looking for a breakthrough idea. At this stage, you don't worry about how to implement the idea. That would stifle your creativity. Your energy is focused on the breakthrough first, which becomes the beginning of your solution.

EXAMPLE

We want our managers to take their budgets seriously, don't we? Here is one conflict diagram that describes why the disease was never cured.

We want to have An Effective Measurement System. See Figure 5.

An Effective Measurement system is the "opposite" of our disease or core problem. A good objective signifies that if we attain it, we have overcome our core problem. In this case, we pick "An effective measurement system" as our objective.

We fill in two requirements to meet that objective, and the prerequisites which are in conflict. We read it back to ourselves as follows:

In order to have an effective measurement system, **we must have** measurements that improve a given area or function.

In order to have measurements, **we must** measure managers based on things that they directly control (local optima).

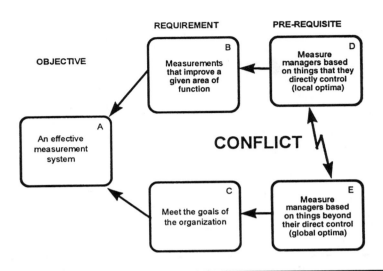

Figure 5   Conflict Diagram

> **In order to have** an effective measurement system, **we must** meet the goals of the organization.

> **In order to** meet the goals of the organization, **we must** measure managers based on things beyond their direct control (global optima).

Think of global optima as those that meet the needs of the overall organization (e.g., profit, throughput, return on investment). Think of local optima as those that meet the needs of a department or function (e.g., production efficiency, performance to budget, product margin).

Each connection between the boxes in the cloud is read as above. The conflict between the prerequisite boxes (the two boxes on the extreme right) are read: On the one hand, we must measure managers based on things that they directly control. On the other hand, we must measure managers based on things beyond their direct control. We cannot do both.

This diagram has many assumptions under every arrow. For example, under the conflict arrow, we assume that we can never have measurements that meet both local objectives (those that are under our control) and global objectives (those of the overall organization). Why not? One assumption is because such a measurement is too complicated for people to understand.

If we find an assumption that is erroneous, we can break out of our cloud and start to solve the problem. This is called an injection, or starting idea. An injection that would overcome the assumption above is "We have a set of easy to understand measurements that link local and global optima."

Constructing a Good Solution — The Future Reality Tree

There is no shortage of good ideas — just good solutions. What is the minimum effort needed to cure the disease? That is the question that this Thinking Process answers for us.

There are three essential elements of any good solution:

1. You can begin implementing in days, weeks or months — not years.
2. It will result in some measurable, significant improvement to the organization in days, weeks or months, not years.
3. There are no, or only negligible, negative side effects.

The conflict diagram provides a starting idea. However, by itself, one idea is usually not enough to cure the disease. We need to create the environment where, instead of these undesirable effects, their "opposites," the corresponding desirable effects, will exist.

The process of building the Future Reality Tree leads to the missing elements –what additional ideas are needed in order to reach our desired outcome. Bearing in mind that all too often a brilliant idea turns sour, we also carefully examine the proposal to ensure that the solution will not cause new, devastating side effects. These additional efforts complete the solution, the set of things that should be injected into our environment.

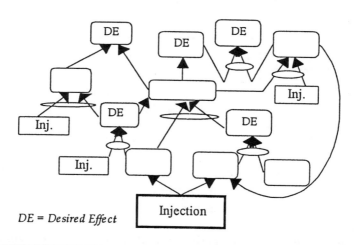

DE = *Desired Effect*

Figure 6   The Future Reality Tree

The Future Reality Tree underwrites a good solution, through three skills:

1. **Ability to determine the characteristics of a good solution.**
    What "desirable effects" must we witness to assure us that we have
    significantly improved our organization? It's not obvious. For exam-
    ple, just because "managers are leaving" is undesirable, it doesn't
    mean that "managers never leave" is the desirable effect. Often, desir-
    able and undesirable effects are not opposites.
2. **Ability to construct the solution strategically.**
    We use the identical rules of logic to verify that the ideas or injections
    will lead to the desired results. The logical structure of the Current
    Reality Tree helps this process move very quickly.
3. **Ability to eliminate all significant negative side effects.**
    Every idea has negative effects as well as positives. At this point, we
    can be so caught up in our solution, that we forget to think about
    the medicinal taste it might leave in other people's mouths. We want
    no harm to come to the organization from the solution. This tech-
    nique ensures a positive result.

EXAMPLE:

We begin by reconsidering the undesirable effects in our current reality and
thinking about which ones must be eliminated. We then determine what
desirable effects should replace them. Here are two examples:

Undesirable Effects	Desirable Effects
■ Customers are not buying as much as they used to.	■ Customers buy greater quantities more frequently.
■ Demands on customer service are going through the roof.	■ Customers are quickly serviced by friendly staff.

All of the rectangular boxes (Figure 7) are the injections or ideas that will
cause our solution to happen. At this point, we don't worry about how to
implement the ideas. That comes next.

> *Using the rules of logic, we make sure that the solution makes sense to
> everyone involved, not just the inventor.*

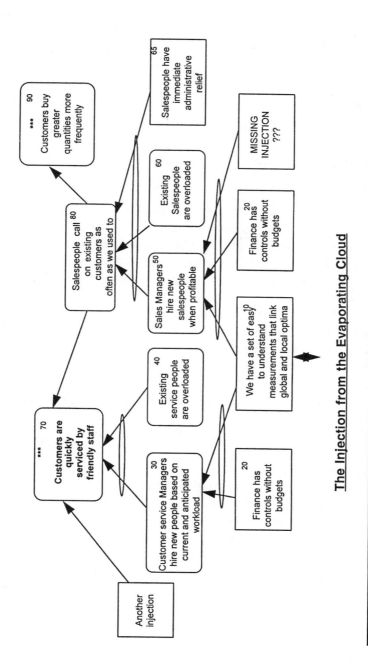

The Injection from the Evaporating Cloud

Figure 7   The Future Reality Tree

In this case, one of the people involved did not think that the ideas at the bottom of the diagram were enough to cause sales managers to hire new salespeople. She thought something was missing. Perhaps the missing injection might read "Sales Managers are trained and thoroughly understand the new measurement system." Or it might read, "Sales Managers are not afraid of the new measurement system." Once the people responsible for the strategic solution agree that this new idea is necessary, at a minimum, to guarantee the results, the idea would be added to the Future Reality Tree.

Team Commitment to a Strong Plan — The Prerequisite Tree

We must take all of our ideas from the Future Reality Tree and implement them. That requires team effort and special projects. Over 90% of projects finish either late, over budget, or not within the original scope. Often, the problems that cause a project or team to fail can be identified early; however, at the beginning of a project, team leaders are very busy and tend to suppress or avoid negative comments.

Unfortunately, unlike the good old days, most people on today's teams are part-timers. They have a lot of other responsibilities and they work a lot of hours. The easiest excuse for not getting something done is "I have other commitments". To address this problem, the Prerequisite Tree requires leaders to use people's negative fears and obstacles in a very positive way. In this way, the team will complete the work earlier than expected, with far less frustration.

Using the Prerequisite Tree, an effective team plan:

- Recognizes obstacles.
- Sets milestones (objectives) to overcome obstacles.
- Synchronizes everyone's efforts into a logical sequence of events that everyone agrees to in advance.

The Prerequisite Tree builds vital team commitment. The result is more realistic, stronger plans, even for teams of 1, through three skills:

1. **Ability to prioritize the significant hurdles to implementation.**
   Every team member has their own intuition about the effort required to implement an idea and potential snags. If problems go unrecognized, team members shift their priorities to other projects. Just like

in the Olympic hurdles, team members must see how many major obstacles there are and how far it is to the next hurdle, in order to get their commitment to run.

2. **Ability to translate hurdles into acceptable conditions.**

Once a hurdle is recognized, we need to envision the condition that tells us that we've overcome the hurdle. Not all hurdles need to be eliminated. We can get around some of them with far less effort than it would take to eliminate them.

For example, consider a project to eliminate smoking in a manufacturing plant. A team member says that a significant barrier to completing the project is "Smokers are not allowed enough breaks." The problem that they envision is that some people are physically addicted and without a break every hour, they will continue to smoke on the job. One acceptable condition that tells us that we've overcome this problem might be that "Smokers are coping with addictions without impacting productivity." The acceptable conditions become the project milestones.

3. **Ability to get group consensus on sequenced milestones.**

Busy people hate to do work that isn't immediately necessary. Projects can be sequenced in many different ways. The correct sequencing can save a project's life.

With this skill, teams examine each milestone step to determine where it best fits. Discussion centers on the assumptions behind why one step must follow or come before another. By learning how to surface the assumptions behind a given sequence, team members build a plan that makes sense to every team member.

The Prerequisite Tree is completed for each idea (injection) or set of ideas in a Future Reality Tree that is complex to implement and will benefit from stepping stones. When complete, the diagram appears as in Figure 8.

The milestones are intermediate stepping stones or objectives that signify that we have overcome the obstacles.

EXAMPLE:

We've decided to implement our first idea that came out of our evaporating cloud. The idea is that "we have a set of easy to understand measurements that link global and local optima." There are just a few small obstacles to overcome before this idea can be implemented, e.g.,

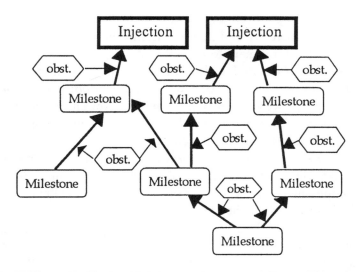

Figure 8   The Prerequisite Tree

Obstacles That Block Us from Having a Set of Measurements

- We don't know what global optima to use.
- We don't know how to link global and local optima.
- Some managers may not like or understand the new measurements.
- Managers may lose bonus during the change over.

Once everyone agrees on the hurdles — we must now turn them into acceptable conditions that signify that we've overcome the obstacles. Each Acceptable Condition is a milestone:

Hurdles	*Acceptable Conditions/Milestones*
1. We don't know what global optima to use	1. A set of global optima is accepted by all senior management.
2. We don't know how to link global and local optima.	2. We have several acceptable examples of how to link global and local optima.
3. Some managers may not like or understand the new measurements.	3. All managers fully accept the new measurements.
4. Managers may lose bonus while changing over.	4. Managers do not suffer financially during the transition period.

Our final step is to sequence the milestones into a plan that makes sense to the entire group. For example, one team member reasons that "*before* we can have all managers fully accept the new measurements, we *must have* several examples of how to link global and local optima." Another team member argues that you could have managers accept the measurements first.

The first team member responds: "The reason we need the examples first is because managers will not accept something if they don't know how to do it". Surfacing of assumptions behind the sequencing is an essential activity to determine a sequence that every team member can live with. In fact, I don't know of any better way to gain 100% buy-in to the plan.

Our team is now committed to a logically sequenced, coherent plan.

Implementing Change Successfully — The Transition Tree

To finally secure the future, you must successfully reach every milestone you have set and reach it on time.

Many great ideas fail in the implementation. To prevent such failure, we must ensure that we're taking the *right* actions. Also, we must check that the actions are *sufficient* and *in the right sequence*. The Transition Tree provides a powerful tool to support delegation and to meet difficult milestones.

This method forces those implementing a solution to carefully examine the minimal actions really needed to *ensure* the required change. But above all, putting *gradual* change as the backbone of an action plan provides a safety net which is essential when planning the future. Simply, the important thing becomes *causing a specific change in reality*, rather than sticking to a specific action just because it is planned.

By teaching people how to overcome impending disaster in every action plan, the Transition Tree significantly reduces and, over time, eliminates failures in implementation and unnecessary rework. The Transition Tree ensures successful implementation of change, through three skills:

1. **Ability to identify the *correct actions* required to cause change.**
    How can you tell, in advance, whether or not the actions you plan to take will work? A scientist, according to international standards, rigorously plots an experiment — each action they will take, each intermediate result they expect, and the end goal. In science, each intermediate result is expected to be verifiable and repeatable. A good

scientist combines their knowledge with the knowledge of others. By doing so, they increase the probability that the actions required to achieve results are correct. We don't always need the rigor of a scientist to turn every milestone into an achievement. Some milestones, however, are a tremendous challenge.

2. **Ability to identify the *evolution of results* required to meet a milestone.**

   Suppose you are writing a novel. You are now describing the part where the executives of a billion dollar company attend a meeting to approve the hero's dramatic, new solution. How would you make the story of the meeting unfold? A *series* of results must occur before the milestone is reached. This skill ensures that participants know how to identify the evolution of results towards the milestone. This same skill is used to verify that a new procedure will work.

3. **Ability to ensure actions are sufficient to meet a milestone.**

   What makes you so sure that a particular action, by itself, will get the desired result? The only way to test this is to be able to identify the assumptions that claim that this action, by itself, is good enough to get the result. As you bring out these assumptions, you can quickly discover where your actions are not good enough. The Transition Tree also allows you to easily correct the situation, by adding actions and additional stepping stones.

A complete Transition Tree is read as follows (Figure 9): I must address need 1, so I take Action 1, with the objective of getting Result 1. Why am I so convinced that I will get Result 1? Because of Assumption 1. If I get Result 1 and a condition exists, then this leads to Need 2. etc.

EXAMPLE:

Your VP of Human Resources delegates, to her secretary, the goal of having all managers accept the new measurements. The VP wants everyone in the country to sign a written agreement to that effect. The VP briefly outlines the four actions needed to complete the task. She expects this task to take two weeks, but months later, the task is still not complete. In spite of weekly reviews, no progress is being made.

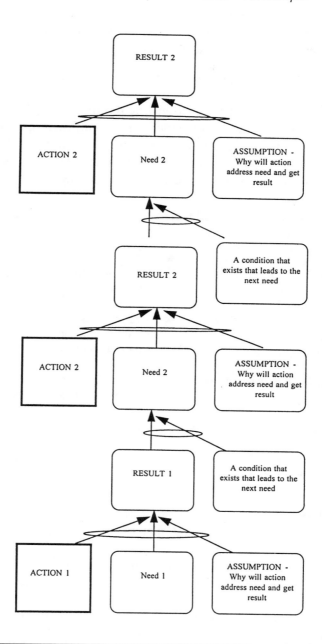

Figure 9   The Transition Tree

- Explain the new measurements by telephone and E-mail.
- Give every manager a one-week deadline to sign off.
- Get everyone's questions answered.
- Tell anyone who is late that the President has mandated the change.

As the VP reviews the list of actions she recommended, she is troubled. She realizes that the actions are neither correct, sufficient, nor in the right sequence. Using the Transition Tree, she puts the actions into a form of intermediate results and assumptions. The four actions turn into forty! For example, before any manager would sign off, the following steps are a must:

Actions	Results
■ Design a training course to explain the purpose and the mechanics of each measurement	■ We have detailed, written documentation that explains in simple terms how the new measurements work.
■ The boss of each manager informs them they are enrolled in a training program.	■ Every manager is committed to attend a training program on the new measurements.
■ Managers resolve any troublesome issues with their boss.	■ Managers have resolved all outstanding issues with their boss or have a meeting with the President to resolve other issues.
■ The boss has the manager sign off.	■ Every manager, except those going to see the President, has signed off.
■ Etc.	■ Etc.

The VP then reviews the planned actions with others, and discovers more errors. The result is that a project that made no progress for months is successfully completed two weeks later.

Figure 9 shows what a small portion of the Transition Tree would look like. A partial reading of this tree sounds like a storybook unfolding:

> **IF** new measurements need to be explained, **AND** Joe writes a manual that explains how the new measurements work, **THEN** we have detailed documentation that explains the new measurements in simple terms within a week **BECAUSE** Joe is the best writer we have. He is available and understands the intent of the new measurements.

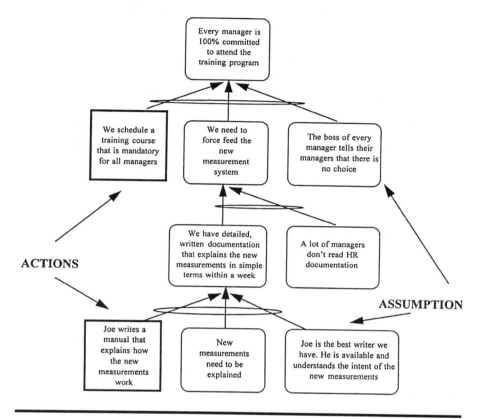

Figure 10   A Small Portion of the Transition Tree

As a test of whether or not you can gain value from the rigor of using a Transition Tree, consider a piece of data and a suggestion:

> *People who use the tool claim that over 70% of their actions are different than the ones they would have taken without the tool. They also believe that they would have failed entirely or taken much longer to achieve the result without it.*

In summary, the five Thinking Processes of the Theory of Constraints, and their purposes are:

■ *Current Reality Tree* — A process to separate symptoms from their underlying causes and identify a core problem — the focus of our improvement efforts.

- **Conflict Resolution Diagram (Evaporating Cloud)** — A technique that shows why the core problem was never solved, and fosters a new, breakthrough idea.
- **Future Reality Tree** — The strategic solution to our core problem, identifying the minimum projects and ideas necessary to cause improvement.
- **Prerequisite Tree** — The detailed plan of all the obstacles we need to overcome to implement the ideas and projects in our Future Reality Tree
- **Transition Tree** — The actions we need to take, and why, to fulfill our plan.

In the next chapter, we review other assumptions and concepts behind the Theory of Constraints.

4 The Assumptions Behind the Theory of Constraints

Now that you have seen an example of the Theory of Constraints in action, we will examine more of the assumptions and thinking behind the Theory of Constraints. We are particularly interested in assumptions that impact our ability to secure the future of our organization. In order to secure the future, you must remove diseases (core problems) from your organization, not just symptoms. In fact, you must remove diseases to ensure two things:

- That you meet the necessary conditions of long-term existence of the organization.
- That you remove blockages to improvement — open the clogged arteries.

An assumption of the Theory of Constraints is that in every organization, dependencies exist between functions, processes and people. An organization is a system. A system is like a chain with interconnected links. And a chain is only as good as its weakest link.

How many weakest links are there in a physical chain? **One.**

However, an organization is not just one chain, but several. Every set of interdependent events or functions or processes leading to a measurable outcome for the organization is a chain. For example, manufacturing a product requires a chain of events. Getting a customer to buy that product requires a chain of events.

Many people confuse chains with processes. This becomes even more confusing in the light of "Process Management" consulting that is becoming more and more popular.

A chain refers to the entire set of interdependent events that must occur to achieve a result. Therefore, in Theory of Constraints terms, a chain may refer to several "processes" that interact. For example, to move from receiving a customer order to producing a product and getting cash back in our bank account from selling the product, we must look at a wide range of interdependent events. The events actually begin with our ordering goods from our suppliers and may not end until a retail consumer buys the good from someone that we sell to.

In this overall process, engineering may be involved in some design work, manufacturing is involved in production, sales and marketing are involved in finding the customer and keeping the customer and internal people informed, accounting is involved in tracking the business transaction, the warehouse is involved in handling and storing the finished goods prior to shipment, a transport company handles the shipping, etc.

How many chains are there in an organization? The Scarborough Public Utilities case study provides a good example of this. The answer depends on the organization, but it is not unusual to find dozens of chains, even in a relatively small organization.

Let's look at the concept of dependencies in a real life example. Consider the example situation described in Chapter 3, where company sales were stagnant, profits were dropping and management turnover in the company began increasing. The Finance department has a policy that no manager is allowed to go over their budget. The Human Resource department puts a freeze on hiring. The Sales VP, who gets measured on performance to budget and tries to follow company policies, stops hiring new salespeople and managers. Existing customers cut orders way back. Why? Because the salespeople, who used to service the existing accounts, are so overwhelmed with work that they no longer have time to call on existing customers. Sales managers are leaving the company in frustration, without the necessary and justifiable manpower to get the job done. What is the weakest link? That is what the Theory of Constraints is designed to identify and eliminate.

Good strategies must recognize these dependencies. Rather than attacking the symptoms of a problem as they exist within any one department or process, we must attack the very root of the disease. This requires analyzing and making connections between processes, between functions, between levels and departments. This often requires the courage to kill some sacred cows across the entire organization (such as measuring managers on performance to budget or on variances from standard costs).

Another assumption lies with a concept of statistical fluctuations and chaos theory. No matter how good our forecasting systems get, no matter how good a manager is at getting employees to show up at work on time, things beyond our control will happen. Therefore, any good, practical solution must accept that chaos happens and not rely on perfection, even from the blessed computer.

The concepts of dependencies and statistical fluctuations are illustrated beautifully in Dr. Goldratt's bestseller, *The Goal.*

There are many other assumptions that we'll discuss as we go along. But that's enough to get us started. What you need to be aware of is that every theory has assumptions. Assumptions are open to challenge. They may be universally valid or invalid, or valid only for a given set of circumstances. They may be valid today and be the very thing that kills us tomorrow.

In order to have a secure future, you must be open to continuously verbalizing and examining your assumptions and having them challenged. You must also have an atmosphere in your organization where people can state their assumptions without fear of being ridiculed.

Physical and Invisible Constraints

Every organization deals with two types of constraints — the physical and the invisible.

Physical constraints are those that you can see and touch. They are typically physical resources, such as a machine which can not produce enough product to meet demand or insufficient numbers of people to finish a task by a given time. Physical constraints are often much easier to identify and resolve than invisible constraints. For example, if you have a physical constraint in a factory, you can often spot it by the huge amount of inventory waiting in front of the constraint to be processed.

Examples of invisible constraints are policies, training and measurements — those things that we cannot see but nevertheless destroy our organizations. Policies, skills and measurements, which probably made sense at one time, are slowly killing most organizations. When a policy, skill or measurement causes people to take action that loses money for the company, it has become a disease. Yet many diseases go unrecognized. As a result, managers implement strategies with no lasting impact, because the strategies do not remove the disease permanently.

In applying the *Theory of Constraints* to a complex problem, we often find that the root problem is not a physical constraint, such as a bottleneck in our distribution system or a manufacturing plant or head office. It's often measurements or policies that cause managers to strive for things that are good for their department or function. However, what's good for the department or function frequently is in conflict with what's good for the overall organization. This very situation was described in Chapter 3 in our example conflict resolution diagram on global vs. local optima.

> *Some conflicts, if they are not resolved, eventually kill the company. These slow killer diseases are not recognized until it is too late. Things seem to go downhill quickly, when in fact the diseases had been around for years.*

Removing the Constraints to a Secure Future

I suggest that a sure way to secure the future is to think systematically. This implies three attributes:

- You have identified changes that will cure diseases across the organization.
- You have great confidence, *a priori,* that the changes you plan to make will work.
- You have the buy-in of everyone involved in the changes.

The Thinking processes of the Theory of Constraints and their objectives are designed to get major improvement, by correctly answering the three questions of change:

STEP 1:   What to change

This step identifies the constraint that must be removed to secure the future. The following Theory of Constraints analysis process can be used to answer this question:

- **Current Reality Tree** — Find the diseases or core problems which are causing most or all of the symptoms. Using a combination of intuition and cause–effect logic, you understand the interdependencies that exist between functions, people, and processes.

STEP 2:   What to change to

This step identifies the elements of a strategic solution — the ideas and the resulting positive effects that must exist for the solution to be practical and acceptable. The following Theory of Constraints analysis processes can be used to answer this question:

- **Conflict Resolution Diagram** — Recognize and analyze the paralysis that's prevented us from attacking organizational diseases. From the analysis, we also derive some starting ideas for a solution.
- **Future Reality Tree and Negative Branch Reservations** — Define the strategic elements of a solution — the main ideas — to determine if these ideas will lead to the effects we want, and to eliminate negative side effects (where the cure may be worse than the disease).

STEP 3:   How to make the change happen

This step identifies the obstacles that must be overcome to implement each idea, and the minimal actions required to implement. The following Theory of Constraints analysis processes can be used to answer this question:

- **Prerequisite Tree** — Take a strategic idea and break it down into milestones, based on the obstacles to implementing the idea.
- **Transition Tree** — Take a milestone or idea and determine the correct actions and sequence of action steps required to implement the idea.

Can Two People Both Be Right?

No matter what methodology we use, we're looking for answers to address issues of change. A good, practical solution is one which can be implemented and which, when implemented, will lead to improvement, moving us closer to our goal. However, even good, practical solutions meet enormous resistance in organizations.

In most organizations that I visit, I see how difficult it is to gain agreement among the management team. There are many conflicts, even when we agree on the goal. For example, there is a conflict between short term and long term strategies. Many organizations ignore the dependencies between making money in the future and customer satisfaction and between employee security and satisfaction and making money, and between employee satisfaction and customer satisfaction. As a result, strategies go awry and management dis-agree with each other.

As an example, several years ago, I was charged with implementing the sales force automation program for over 200 field people. The VP of Oper-ations was not supportive, because this project was taking a million dollars away from much needed operational improvements. However, the President, a technology "nut" and totally in favor of the automation project, overrode the VP and the project was implemented. With the lack of buy-in from the VP of Operations, the laptop computers were delivered to the field sales force without inventory lookup or on-line order entry — two key elements which were necessary to gain company benefits from the project and secure a competitive advantage.

In fact, this company desperately needed to undertake several major projects, recognizing the dependencies between their results, in order to improve. However, with the company in a money-losing situation, the Pres-ident had no way of either recognizing or presenting such a strategy to the already impatient board of directors.

If senior executives really want to focus energy on finding and imple-menting good, practical solutions, rather than arguing, they have a much better chance at success using a process and a set of rules. The process should be a series of steps that they can follow in coming to *logical rather than violent agreement*. The rules should add clarity and value in carrying out the process.

> *Everyone, at any level in the company, must be allowed to logically question important things that do not make sense to them, if they are impacted in any way.*

The Theory of Constraints provides eight such rules of logic to analyze existing situations and to predetermine whether or not a future strategy will achieve desired results. See the Glossary for a complete description of the Eight Rules of Logic (Categories of Legitimate Reservations).

Imagine, in a family situation, a husband or wife saying, "In order to earn a good living, I'm going to take a substantial bank loan and buy a brand new, white corvette convertible." Is this nonsense? Has the person gone mad? Rather than blowing up with high emotion, let's use a few rules of logic to question this statement.

The first rule we might apply is one of *clarity*. What is it that this person means by "a good living." Let's suppose the person responds by saying, "A good living, to me, is $100,000 per year." OK, that does clear up my confusion.

Let's apply another rule — *causality.* We do not understand how buying a corvette convertible will *cause* them to earn a good living. The person responds, "I have a two-year signed contract with a Fortune 100 company executive that retains me for $100,000 per year to be available 12 hours per day, 5 days per week to chauffeur her and her clients around town in a white corvette convertible."

OK, but now we have a concern about negative side effects if we take this action. In Theory of Constraints terminology, this is called a Negative Branch Reservation. This is not a rule of logic. It is a subset of the Future Reality Tree, expressing logically how the implementation of our idea leads to some undesirable negative consequences.

Our concern is, what happens to this investment if the contract is not renewed in two years. The person responds with a published document from a respected source showing how well-maintained Corvettes hold up to 90% of their value over the first four years in operation.

This kind of discussion could have been disastrous. The fact that it proceeds constructively is not by accident. Furthermore, this example is mild compared to some wild ideas that I've been exposed to at senior executive and middle management levels. This strategy affects one family. Imagine a strategy that could impact hundreds or thousands of people and their families!

Is this the best strategy? Who knows and who cares!! What we want is a good strategy — let's not waste time trying to find the "best" strategy. I'll leave "optimization" to university professors and opt, instead, for practicality. As long as we know, *a priori* (in advance) that our strategy will move us closer to our goal, that's what counts.

Using logic and rules, we can better determine whether any individual strategy is likely to succeed. Using common measurements and criteria, we can compare the impact of different strategies on profit and return on investment.

To secure the future, however, we must recognize that necessary conditions are goals that must be achieved. They are as important as any primary goal we may set. For example, I suggest that strategies which focus only on cost savings and ignore the need to increase Throughput or sales are not strategies at all but rather short-term survival measures. Strategies which dwell on intangibles, such as employee or customer satisfaction, and fail to quantify expected benefits are avoiding an understanding of the cause effect relationships required to get results. For example, if we focus on quality improvement in the door handles on cars to get them to within a tolerance of one thousandth of an inch (as one TQM guru advocated), will that result in more sales?

We contend that, if an organization is already heading in the "right" direction, a good strategy is one which removes obstacles to improvement. By finding the weakest links and strengthening them, we are implementing a good strategy.

What about organizations that are not heading in the "right" direction? For example, there are organizations with products which are at the end of their life cycle, or in sunset industries. Those organizations must do a market analysis to understand and come up with solutions to their client's underlying problems — solutions which will give them a long-term strategic advantage. Or they may want to address new markets, in which case a market analysis of the new market is required.

Once again, a good strategy in this situation is one that the entire executive team and the entire company would buy into, because it is based on logic combined with an in-depth analysis of customer needs. If we cannot prove internally, in our own organization, how a series of actions, services and products will provide significant value to our clients, how could we ever possibly hope to convince our clients or our bankers?

Here is a final thought about what the Theory of Constraints advises about moving into new markets or new product areas. If employment security and satisfaction is a necessary condition to our long term security, then new markets and new products should use existing resources.

Conclusion

Finding a good set of strategies is not a matter of luck. Nor is common agreement among our management team sufficient to succeed. If we define a good strategy as one which removes obstacles to improvement, then we must figure out how to identify and remove those obstacles. If our current direction is not good, and we decide we must move in a new direction in order to improve, then we must identify those underlying customer needs that will provide a long-term strategic advantage in our target markets.

Pick either approach or both. We still must have a thinking process or methodology to correctly answer the three major questions of change — what to change, to what to change and how to make the change happen. Answering these questions logically means that the solution is common sense to everyone involved. The ability to construct and communicate common sense through logic is essential to develop and gain buy-in to strategies.

5 Three Diseases Attacking Every Organization — Policies, Training, and Measurements

The purpose of this chapter is to familiarize you with three of the most common organizational diseases known to man — policies, training, and measurements. Chapter 17 provides an extensive review of measurements and the characteristics of successful measurements. The Case Studies also document several examples of core problems relating to measurements and how they were solved.

I can't wait until a favorite competitor opens their huge home store across the street from where I shop today. Let's call the place where I shop today store X. I love the store contents. Store X has items that I cannot find within 50 miles. But *every* time I shop there, it costs me at least an hour of my time. Merchandise is often unmarked. The staff sends me on a wild goose chase to find things. They rarely have complete sets of merchandise. Eventually, if these problems go uncorrected and a competitor does just a little bit better, store X will be out of business.

When store X dies, it will be poor policies, training, and measurements that formed the death sentence. It's a slow death, and often goes unnoticed by executives and staff alike, until it's too late.

It would be a rare patient who came to a doctor saying, "I have a brain aneurysm. That's why I have headaches, blurred vision, high blood pressure, etc." More typically, the patient gives the doctor the symptoms. The doctor speculates on what might be causing those symptoms (effects). They may even run some lab tests. Eventually, through a combination of intuition, experience and cause–effect logic, the doctor arrives at a diagnosis — the disease causing the problems.

For people to have success in improving organizations, they too must recognize the difference between the symptoms of a problem and the disease.

> *One important factor that enables doctors to come up with a diagnosis is that they are already familiar with diseases — they've been trained in diseases and their resulting symptoms. Managers have not.*

Measurements

In Chapter 3, I provided an example of the symptom of a disease — merchandise in a store not marked with a bar code symbol. Look at all of the other negative symptoms that arise from this one — waste of time for all customers in that cashier's line, unhappy customers, waste of cashier time, waste of floor staff time, probably a waste of the manager's time, bad morale of employees who do a good job of ticketing merchandise but have to listen to the manager's speech. Previously, in Chapter 3, I indicated how management would likely treat the symptom and fail to cure the disease.

Dr. Goldratt wrote in his book, *The Haystack Syndrome*,* "Tell me how you measure me and I will tell you how I behave". My converse to this is "Tell me how you're not going to measure me, and I'll tell you how I won't behave."

Chances are many store employees frequently encounter unmarked store merchandise. If it is not within their department or job function, what do they do? They ignore it! In this store, no single employee suffers as an individual if merchandise is not marked. Can we speculate that the disease is a measurement that is either lacking or invalid?

If this is true, then a good measurement, in combination with other elements of a practical solution, should fix this problem permanently and eliminate most or all of the related symptoms. Suppose, for example, that we implement a new measurement as follows:

> Every week, a pool of $1,000 will be available to be shared by all employees, in proportion to hours worked. To simplify the example, assume that we have 25 employees in the store, and so each one can add $40 to their weekly paycheck. Is it a significant enough incentive to capture their attention? You bet!!

---

* *The Haystack Syndrome*, Eliyahu M. Goldratt, 1990, North River Press.

For every incident of unmarked merchandise that is identified by the cashier, $50 gets taken out of the bonus pool. Let's assume that cashiers are on a different kind of bonus, so that they still have the incentive to report such problems.

With this kind of measurement in place, is it more or less likely that every employee in the store will look for unmarked merchandise and correct the problem promptly? If it's truly the measurement that is the disease, and this is a good, practical measurement, then clearly it should reduce, if not eliminate, the problem of unmarked merchandise in the store.

CAUTION: **Measurements may be dangerous to your health!!**

Before implementing any new measurements, remember that every new idea has some negative side effects. Therefore, take any proposed new measurements and then:

- Review them with the people who will be impacted.
- Remove all serious negative side effects. To the extent possible, remove *all* negative side effects pre-identified.
- Pilot before a full implementation. Note that piloting does not imply coexistence of old and new measurements for the same people. See the Acme Manufacturing case study for their experience and recommendations.
- Keep the people affected by the new measurements involved in the pilot, taking all of their obstacles into account before finalizing a new measurement.

Measurements can also cause problems because they are outdated. They may have worked well at one time, and therefore are considered to be "Proven" by the management team or the Human Resources department. However, things change. The good reason for the measurement, several months or years ago, may be gone. The result may be that this once valid measurement is causing all kinds of pain and bad behavior problems.

For example, think of an efficiency measurement in a manufacturing plant, that causes workers to keep running a machine and building up inventory, when there is absolutely no short term need for the inventory. Think of a sales force that gets measured on the amount of margin they bring in rather than total throughput. Is it possible that they'll try to sell the most "profitable" products, regardless of the company capacity to produce or procure those products and regardless of what the customer really needs? Think of managers that get measured on performance to budget. Are they likely to add much needed headcount if it will make them go over their budget? The examples in every single organization are widespread.

How do you identify that it's an erroneous or missing measurement that's causing the problem? By using cause effect logic, you can look at the symptoms and work your way down to the root cause. This is the Current Reality Tree.

Policies

How many times have you heard these poisonous words? "*I'm sorry. I can't do anything about that. It's our policy*". Whatever happened to all of this so-called empowerment that people have? In reality, most organizations empower people with responsibility but rarely balance that with the authority they need to resolve problems and avoid aggravating other employees, customers and suppliers.

The other major problem with most empowerment programs is that they give certain employees the power to override policies without training them or examining why the policiy exists in the first place, i.e., what company interest is the policy supposed to protect?

Here are just a few examples:

A multi-billion dollar steel producer thinks that they have a physical constraint. Their caster cannot process enough steel to meet customer demand. Of course, with this symptom are many other symptoms of a disease. However, there are some facts that indicate that they have misdiagnosed. For example, they have excess inventory at all levels — excess raw materials, excess work-in-process, excess finished goods. Generally speaking, if you have excess inventory at all levels, then you have excess capacity — it is not a physical constraint that's blocking you from making more money.

Let's put some other symptoms together with these, and see how this leads us to a policy causing all of the related symptoms. Another symptom is that some customer orders are finished early, leading to excess finished goods, while other customer orders are finished late. Some of the late orders won't get used for 9 to 12 months, because we've missed the customer's production cycle for that steel.

For example: a steel company supplies the metal for a customer that cans tuna. The tuna company goes through a canning process every 6 months. The tuna are sitting on the dock, waiting for the cans to arrive, but because steel producer A is late, the company uses steel producer B. After all, you can't leave hundreds of tons of tuna sitting around indefinitely. The tuna company has a contract with steel producer A, so they definitely must take delivery of the steel at some point. They tell steel producer A to keep the steel for another 6 months until they need it the next time.

Why do we finish some orders early and others late? Because we schedule like orders together. Orders are grouped according to grades and width of steel required and other parameters. Why does a steel company schedule this way? If you guessed "Policy," backed up heavily by "Measurements," you are correct.

Steel companies have a 100-year history of measuring themselves by the number of tons of steel they produce. While this may have been a good measurement when there were few grades of steel and manufacturing was less sophisticated, today this type of measurement kills smokestack industries.

When you go through all of the layers of cause and effect, what you find is an astonishing fact. This company has 100 groupings — categories of steel based on combinations of thickness and metallurgical properties. Do they need them all in order to manufacture a satisfactory product for their clients? Absolutely not. You could produce with far less excess inventory and far less cost with less groupings. Why do they have the groupings? Policy — The company has had a policy for many, many years to sell steel according to the various grades. True, they must guarantee certain minimum grades to certain clients. A major reason for the other grade groupings is for pricing consid-erations.

Now picture this scenario: I sell two types of drills. One is for the home market, and one is for the professional market. The products look absolutely identical, but the professional drill is stronger inside — it has a better coil, stronger bits and better quality wire. The professional drill costs $10 to manufacture and sells for $60 to retail chains. The home drill costs $6 to manufacture and sells for $25 to retail chains. The professional market demands a high-quality, continuous use product with a five-year guarantee. The home market demands a light use reasonable product with a two-year guarantee.

Suppose that I have two manufacturing lines for the two products. The home line is used 3 shifts per day, 7 days per week. I cannot meet all of the demand for the home market drill from the one manufacturing line. The professional line is used 4 hours per day, 5 days per week. What makes more sense? To enforce a policy that I can only produce professional drills on the professional manufacturing line, or to change the policy to allow us to use the excess capacity of the professional manufacturing line to meet the home market demand?

Of course, manufacturing experts would immediately come up with all kinds of "other considerations". Some of these may be valid, some not. Identifying the disease is halfway towards solving it, and erroneous policies account for many modern organizational diseases.

A quick check on whether you're dealing with a bad policy or not is to ask yourself the following questions:

- Do I feel a need to do the opposite of what the policy is telling me to do?
- If I do the opposite of what the policy is telling me to do, will it have a positive result for the organization?

If you answered yes to the above two questions, then you may want to permanently change the policy. However, in the meantime, you'll want a workaround. To find a good workaround, ask yourself (and the company policy owners) the following questions:

- Why was the policy created in the first place? Is there a valid organizational need that must still be protected?
- If you answered "Yes" to the above question, is there a way to meet the organization's need and override the policy at the same time?

Training

Picture this: an organization that services hundreds of thousands of customers gets overwhelmed every month when they send out their bills. They are deluged with customer inquiries on the bills — in fact, at peak times, they get 5 times as many calls as they can handle. A customer who calls and gets put on hold by the automated, but totally stupid voice-mail system is told to either hold on to retain their priority in the queue of waiting customers or to fax their inquiry to a fax number and someone will get back to them as soon as possible.

Every month, hundreds of customers go through waits that exceed ten minutes. Some wait as long as 20 minutes. Every month, a few dozen of these customers contact the elected officials who are commissioners to this organization. What do the commissioners do? They create havoc inside the organization, asking the same questions over and over again, month after month.

On investigation, it turns out that there are lots of staff available in the Customer Service area to handle calls. The problem is twofold — one is that handling billing inquiries is more complex than some of the other customer service work, and second, only a certain number of people in the department are trained for this. This is a training problem.

Picture a different organization with a more positive scenario. You call a telephone number and immediately get switched to automated voice mail. This time, the voice says, "Your call is very important to us. There are three people ahead of you in line waiting to be serviced at this time. We handle an average of 300 calls per hour with 30 operators. Each call takes an average of one and a half minutes to handle. Please stay on the line." This was an actual dialogue of a call to an organization you would not believe — a Florida State government office. What a beautiful contrast to the situation described above.

I've waited in home centers for half an hour for someone to show up to cut some carpet or some wood. I've waited days to talk to someone who was knowledgable enough to handle my inquiry on very basic product or service questions. My conclusion is that many North American companies have vastly undertrained staffs and are courting disaster. A smart competitor will put some of these people out of business overnight.

What's truly disappointing in the training arena are organizations that grew strong and fast, based on skills and training, and have now started to fall apart — fast-food chains with servers who handle money and food with their dirty fingers, automobile manufacturers and long-distance carriers who put polite, yet unknowledgable operators on their 800 lines, computer hardware and software manufacturers who can no longer answer the public's questions about their products.

When you combine this lack of basic skills with the inability of most people to think through issues and use cause–effect logic, the results are not surprising.

American brains are as big and powerful as the Germans, the Japanese, the Chinese. All that any of us have to do to secure our future is to learn how to use them more effectively than the competition.

THE THEORY OF CONSTRAINTS AND OTHER IMPROVEMENT TECHNIQUES

6 Why the Total Quality Approach Fails

A great deal of quality effort is having no impact on many individual customers and a resulting devastating impact on a company's bottom line. We'll use cause–effect logic as illustrated in the Theory of Constraints to examine this problem. How is it that a company can improve their ability to produce good product and still irritate the customer? This chapter explains why this happens frequently and what can be done about it.

In a 1994 survey, nearly 80% of U.S. managers believed that quality would be a fundamental source of competitive advantage by the year 2000.* Yet barely half of Japanese managers said so. The Japanese claimed that the capacity to create new products and businesses came first. What this reflects is that by the year 2000, quality will no longer competitively differentiate; it will simply be the price of market entry.

Quality with Negative Customer Impact

A major North American auto manufacturer boasts that quality is their number one focus. Yet, in the 90's, their market share declined, spelling impending disaster. It's true that their quality has improved immensely, if you examine all of their internal statistics. It's also true that they are ineffective in extending the quality chain to many of the aspects that impact the customer.

---

* *Competing for the Future*, Gary Hamel and C.K. Prahalad. 1994, Harvard Business School Press.

In Chapter 3, we defined a chain as the set of interdependent events, functions, or processes leading to a measurable outcome for the organization. Let's extend the definition of the organization to include the entire supply chain.

For every product, there is an end consumer and a beginning supplier. Quality has an internal context and an external context. Or as one of my clients called it, there is little "q" quality (internal) and big "Q" quality (external). Little "q" quality reflects those considerations we have in our manufacturing process for producing a defect-free good. Big "Q" focuses on quality in terms of end consumer perception.

If our goal is to produce more profit from sales of the product or service, we must look at the entire supply chain in order to determine not only where the constraint is but also how to remove it. Often, the constraint is not in the quality limitations of our production process, but rather in the next step down or up in the supply chain.

A great deal of the money and effort expended on quality management is focused on quality problems that will not improve the bottom line. Here are some real-life examples:

- A large automotive dealer quotes a delivery time of 6 weeks, after checking with the manufacturer's production facility. The dealer was not aware, until the week that delivery was due, that the manufacturer was on shutdown that week. The vehicle is actually delivered 10 weeks after submission of order, with the client incurring the additional cost of $1,000 in car rental expenses. No compensation is made to the client. The client vows that this will be the last product he buys from this manufacturer.

- A client purchases a new, executive-class vehicle from an authorized dealership. The dealership completes the credit application in error and submits it to the manufacturer-owned credit company. The credit company sends its first invoice to the customer, with a motto at the bottom indicating that they want nothing less than total customer satisfaction. The customer, noticing the errors, immediately phones the "customer service" (oxymoron) representative, who refuses to take any action to fix the dealer-caused errors. After all, the dealer is a separate legal entity from both the credit company and the manufacturer, and the client is now dealing with the credit company. The customer must work his way through levels of authority in the credit operation, sending faxes and letters, upon which the errors are fixed.

The customer has spent hours of his time and become extremely agitated in the process. The manufacturer has lost another customer for life.

■ The same manufacturer experiences more problems. An electronic switch which opens and closes the sunroof of a brand new vehicle fails. The switch is not available at the dealer's for several days. (Thank goodness the sun roof was closed when it failed). Sometime late during the day of the repair, the dealer informs the customer that they cannot complete the repair that day because the switch arrived unpainted, and they did not schedule any time in their paint shop. On both occasions, the customer had to leave the car at the dealer's for a full day. The customer encounters five such incidents with different repairs over a 3 month period. In total frustration, he sells the almost new car at a loss of several thousand dollars.

Where to Focus the Quality Effort

To make sense out of a quality effort, we must evaluate the entire process as a chain — one that goes from the initial supplier of "raw materials" to the end user of the product. Remember, a chain is only as good as its weakest link. Every organization has limited resources, and therefore decisions must be made as to where to focus the quality effort.

For example, let's reconsider the automobile manufacturer above. They've spent billions of dollars these past few years to get their level of defects per part down below the 1 per hundred thousand level, and to reduce their manufacturing cycle time to less than 10% of what it was several years earlier.

However, they have not changed their paradigm for selling to increase meaningful customer value. With decreasing market share, they have a huge inventory of unsold vehicles. Therefore, their reduction in manufacturing cycle time, while commendable, was not related to addressing their constraint. And in spite of trying to copy the Japanese, they are still way behind them in cycle time. Therefore, reduced cycle time will not give them the competitive edge.

Automobiles have many parts, which means that the probability of a failure with any given customer is still very high. Therefore, most customers will experience failures and need to interact with a dealer repair service. Further, as customer's cars wear out, they will experience increasing failures as they get closer to their decision time for buying another vehicle.

Let's assume that we must make a choice on where to direct our quality effort (i.e., we don't have the resources to address both internal quality improvement and dealer quality improvement issues). Is the quality effort better expended on customer satisfaction with repair consumption (i.e., at the dealer organization) or on repair prevention (i.e., getting the defect level down from 1 part per 100,000 to 1 part per 200,000)?

In a for-profit organization, the question is best answered by relating it to the goals of the company. A company must make money, now and in the future. However, other conditions are necessary in order for the company to continue to exist — customer satisfaction and employment security and satisfaction.

If we assume that customer satisfaction is not a choice, but is essential to our long term business, we must ask a very basic question. Do we have even a faint hope of satisfying the majority of our customers in the near-term with the dealer situations as described above? If the automotive manufacturer, at this stage in their quality effort, channeled their quality funds and resources towards the dealer and the credit company to make life easier for the customer, what would the impact be?

The point is not that the correct answer falls in one direction or the other. The point is that organizations must think about where the constraint blocking improvement is, and therefore where the focus should be. Most organizations have far too many projects on the go, creating a serious quality of life problem for employees.

How would you like to be busting your behind addressing quality issues for that automotive manufacturer, only to find that your market share has declined by five points? In other words, customers by the thousands are telling the manufacturer to take their product and put it in unspeakable places, in spite of all of the improvements that were made.

Sadly, an informal survey by this author of several dozen organizations over the past six months has found over 90% with no outstanding evidence of focus on the customer. These organizations have ranged from health care providers to hotels to utilities to manufacturers and retailers.

Yes, it's only common sense that a major bank credit card company should be able to answer their phones within 10 minutes, especially during the "off-hours". But all too often, this is not done. It would make sense for a telephone company to be able to tell a business, ahead of a move, what their new telephone numbers will be, so that customers can be informed. But a Canadian monopoly refused to do so. And wouldn't you expect a customer service clerk at a retail establishment doing $25,000 in daily business to be able to come up with $75 cash for a refund, especially after a customer has waited in line for 15 minutes?

Why do these situations happen? It must be because, as Mark Twain said, "Common sense is not very common". Why not? The answer is that common sense is not very obvious. This truth, interacting with the fact that people have too many projects to work on, means that people are not thinking about where to focus the quality effort. Another fact of life that impacts the quality effort is the downsizing syndrome. It has taken the desire away from many people to improve their companies. Now we begin to see a host of reasons for the things that continue to drive customers crazy and drive customers away.

Fixing the Problem

The bad news is that there may be no one working on these situations in your company. The good news is that they can be relatively easy to fix, once they are identified.

First, remember that standards like ISO 9000 are not foolproof. While it may be necessary to have your organization up to the standard, in order to be able to sell your product, it does not mean that your ISO 9000 company is of any more value to the customer. If this is true, then the standard, by itself, may only be a necessary condition of doing business without being of any strategic value.

Second, realize that not all customer frustrations are created equal. Therefore, these frustrations must be analyzed in order to determine which ones will impact their "loyalty" or future purchasing decisions. Another way to analyze is to separate symptoms from the disease. You are now a doctor, and your objective is to eliminate the cancer causing all of the symptoms, not just to treat the nausea and headaches. A very effective analysis technique is to relate causes and effects. An example of this is the Current Reality Tree technique of the Theory of Constraints. Fishbone analysis is another popular technique, although it does not incorporate the rules of logic, nor does it have a safety net such as the Future Reality Tree, to verify that your analysis of the current problem is correct.

Often, when examining symptoms vs. the root cause of most of the customer's problems, an organization finds that the disease can be fixed with some policy changes, some training and very little cost implication to the company. This often occurs simultaneously with a major increase in customer satisfaction resulting in major short term and long term gains in profits and competitive advantage. See the Orman Grubb case study for a real-life example.

There is one way to find out what your customers' real beefs are. Ask them! Don't do it just in a survey. A survey won't tell you whether a change will be meaningful to them, or how important that beef is. Therefore, you must understand not only the essence of the beef, but if you were to overcome it, how much more business would you get from that customer. These answers can only come from face to face dialogue, an almost impossible accomplishment in today's world of insane voice-mail systems.

Chapters 10 through 12, as well as several of the case studies, provide detailed examples of how to satisfy customers. Remember, the goal is not Total Quality. The goal is to increase Throughput while simultaneously reducing Operating Expenses and Inventory/Investment. Total Quality is just a means to an end. It contains some assumptions that may not be valid in your environment.

If quality is your constraint, then TQ methods may help a great deal. First and foremost, you must identify your constraint. The next chapter, Chapter 6, describes how to integrate a Theory of Constraints and Total Quality approach.

Conclusion

With limited resources, we must focus our quality efforts where it will have the biggest impact on our organization, both on bottom line and on customer satisfaction. We cannot meet these goals without a direct and deep understanding of the end-user customer, the supply chain and where the constraint is in the chain. This understanding may be the key to our profitability and survival over the long run.

7 Integrating a Total Quality and Theory of Constraints Approach

If quality is not your constraint, why spend any time on it? Is it because it's an overall philosophy? Do we want everyone in the company to think "quality"? Or is total quality simply a necessary condition of doing business today? In my experience, organizations that try to achieve quality everywhere end up with quality nowhere. There are simply not enough hours in the day and therefore no focus and no beneficial results. The philosophy of "quality everywhere" does not direct the limited resources that an organization has to where it counts — addressing the constraint resulting in increased throughput and a more secure future for the company.

For the 20% of companies that have been successful with Total Quality approaches, it's often in one or two areas that you can detect a big improvement directly related to the bottom line. Not coincidentally, for example, a multi-billion dollar steel company claimed to get hundreds of millions of dollars benefit from their Total Quality approach. At the same time, many of the departments that tried to implement the approach failed completely. The reason for the measurable improvement was that the company had a physical constraint — they couldn't produce enough steel to meet the market demand. Every case of rework and scrap was an opportunity for additional throughput to the market. By using quality teams, they successfully identified processes that could be improved to reduce scrap and rework.

The way to integrate the two approaches — Total Quality and Theory of Constraints — to the huge benefit of the organization, is best illustrated first by example. Then we will draw a simple procedure from the example.

In 1995, a manufacturer of springs for tractor-trailer trucks had a physical constraint. They could not manufacture enough springs quickly enough to meet market demand. However, things were changing and they recognized that the market would quickly become their constraint. In order to secure business for next year, they had to produce prototype springs for new, prospective customers, in their factory — the same factory that was physically constrained already.

They really needed to get through 3 to 4 prototypes per month, but were only able to output 2 successful prototypes on average. Every production request for a prototype was in conflict with a revenue-producing production order for a customer. You can imagine how this led to constant battles between sales and production, and between various factions in the factory.

In identifying the ability to produce prototypes as the underlying factor that would impact the market constraint next year, we performed a Theory of Constraints exercise. We used the Prerequisite Tree to determine what obstacles blocked the goal of producing four prototypes per month.

Some obstacles were clearly related to quality issues. When you hear symptoms or obstacles such as, "Engineering and production are working on different specifications" or "There is a lot of rework," then you know you are dealing with a traditional quality issue. As long as the quality issue is part of a constraint, it must be examined. Total quality approaches, such as quality teams, statistical measurements, etc. are warranted. However, the solution is not necessarily in addressing the quality issue directly.

As it turned out in this situation, the company was breaking into the overseas marketplace. Every prototype specification was turning the engineering department on its head. The way that the engineering department typically worked was to try to meet the customer's specification. With the change in marketplace, this was requiring extensive redesign of the springs and extensive retooling. As soon as you are using a different set of tools in a different way, you have to redesign and reschedule the already overworked production lines, to schedule the special tools into the production process.

A total quality approach might focus on statistical processes to make sure that the engineering specifications are correct, that the tools are perfect and that the production workers are properly trained in the use of the new tools. This may have worked eventually, but it would have taken two years to get the plant to a physical capacity where it could add another production line and meet the required prototype activity. By that time, the company would have lost a great deal of business to the competition.

What if you hear symptoms or obstacles like, "Engineering specifications are beyond what the market needs" or "We are not offering the prototype that will best meet the customer's needs. The customer gives us a specification based on how a competitor currently manufactures the product. We can meet the needs of the truck driver without meeting the specification, at much less cost and greater reliability." Then what does Total Quality tell you to do? What if our constraint is invisible — a policy or a measurement, for example. What does Total Quality suggest for these?

The Theory of Constraints provides a set of Thinking Processes to help us determine what makes sense. Total Quality is one tool that makes a great deal of sense where the constraint is quality related. It may make no sense, or actually be our downfall, if we follow it blindly as an overall philosophy. Organizations simply do not have time to put quality everywhere. Workers get skeptical if they must spend time focusing on quality and realize that the company is going down the tubes for another reason.

To integrate the two approaches, we must therefore:

1. Identify the constraint within the supply chain (not just within our organization).

2. If the constraint is a physical constraint, determine whether quality approaches will help in removing the constraint. This would be part of the second step of exploiting the constraint. Use the concept of a quality team to take responsibility for the problem analysis and resolution. Do not let Total Quality blindness or inertia prevent you from seeking alternative approaches which will resolve the problem more quickly or cost effectively.

3. If the constraint is an invisible one, use the Theory of Constraints Thinking Processes to analyze the root problems and formulate a comprehensive solution. Do not let Theory of Constraints blindness or fascination with logic trees prevent you from using a Total Quality approach as an injection into the solution.

In the spring-manufacturing example above, each member of the management team took on a number of intermediate objectives to remove the obstacles to meet the goal. The Total Quality manager had a focus on improving the quality of the communication between functions, which was determined to be a cause of much of the rework.

In our effort to secure the future of our organization, it makes sense not to be a zealot for any one tool or methodology. There are many tools that have validity for a specific set of circumstances. The next chapter looks at other tools, some of which you are probably using, and how they integrate with the Theory of Constraints.

8 The Theory of Constraints and Other Management Techniques

M ost organizations use multiple business improvement techniques — over a dozen, in fact.* You may have questions about how the Theory of Constraints works with, or integrates with some of these techniques. This chapter provides a summarized overview of these tools and how TOC may integrate with them.

Activity-Based Costing or ABC

This is a method to allocate overhead/indirect costs to products or customers. It's purpose is to provide information to view a product or customer's profitability from the point of view of the amount of activity associated with it. Presumably, with this information, we can make better decisions about which products to focus our marketing efforts on, which products to drop, etc.

Dr. Goldratt has highlighted, in numerous presentations and books, the fallacies behind traditional cost allocation as performed by cost accountants. The reasoning is that cost allocations mislead managers to make the wrong decisions. For example, they assume that if a product line is cut, all of the allocated costs go away. In reality, most if not all allocated costs (in the form of overhead and indirect costs) are fixed and therefore do not go away.

---

* *Planning Review,* "Managing the Management Tools," September/October 1994, Published by the Planning Forum.

In reality, many manufacturing plants have some lines which are constrained and others which are not. Furthermore, the constraint to producing more profit may be outside of the manufacturing area completely.

Activity-Based Costing may offer some helpful ideas towards analyzing the activity involved in producing and selling a product and servicing a client. To the extent that you can relate certain activities directly to a single product or customer, you could allocate those costs as direct costs. You could then start to compare throughput per constraint minute where the constraint is no longer a machine but some other resource or process.

The Controller at Acme Manufacturing (see case study) and formerly Manager of Activity-Based Costing, cautions that ABC can easily be misused and become another Standard Cost allocation disaster.

An article describing the relationship between ABC and the Theory of Constraints appeared in the Management Accounting May 1994 issue.* Another excellent article appeared in the January 1995 issue.** You may wish to read them or contact the referenced authors for further information.

Benchmarking

This technique suggests looking for the "best-in-class" for a product or service or process.

Often, a Theory of Constraints analysis of specific benchmark criteria finds benchmarking dangerous, particularly the way it is used by most executives. For example, some executives I worked with recently in the steel industry used benchmarking to find the producer who had the lowest labor cost per ton of product produced. Since their cost was higher than this lowest cost producer, they concluded two things:

- They couldn't hire any more people.
- They needed to focus on reducing their labor costs further.

In North America, in this industry, the proportion of labor cost to total cost is very different than with overseas companies. It is also very different between companies because the industry is very capital intensive, and the state of capital investment is very different between individual companies.

* *Using Activity Analysis to Locate Profitability Drivers,* Charlene Spoede, Emerson O. Henke and Mike Umble, Management Accounting/May 1994.
** "ABC vs. TOC: It's A Matter of Time," Jay S. Holmen, *Management Accounting/*January 1995.

Further, although all companies in the benchmark produced steel, the individual products and mix was totally different. Steel comes in various sizes and grades, and has totally different value to a customer based on these and other parameters, such as lead time from order to delivery. The amount of labor involved also varies with the nature of the products produced.

Benchmarking was causing them to compare apples and oranges. The conclusions from the benchmarking were leading them in the wrong direction. Furthermore, suppose that labor cost is not the constraint. Of what value is the benchmark then?

This is not to imply that all benchmarking has no value. It simply suggests that the logical thinking processes of the Theory of Constraints can help to either validate or invalidate a benchmark and it's relevance to where you should focus your attention.

Reengineering

Business Process Reengineering (BPR) and Downsizing have become synonymous in many business articles. I'm sure that's not what James Champy and Michael Hammer intended in their book on the subject.* John Sculley, then CEO of Apple Computer and Robert E. Allen, CEO of AT&T provided two key endorsements of the book and the methodology. Based on what has happened in those companies, we might question the value of the methodology. AT&T returned –7.4% to investors in 1996. And every restructuring that Apple has completed in the past two years has been followed by declining revenues and huge losses.

However, simply questioning or throwing out the technique would be wrong. The technique can provide order of magnitude improvement in business process productivity to the benefit of the customer. Champy and Hammer offer brilliant insights into fresh approaches at organizing work.

85% of reengineering efforts fail, according to the Economist magazine.** Looking at it from a TOC point of view helps explain why. One of the success stories used as an example in Champy and Hammer's text was for a North American automobile manufacturer. They improved their accounts payable department tremendously, and took hundreds of jobs out of the process.

---

* *Reengineering the Corporation: A Manifesto for Business Revolution,* Michael Hammer and James Champy, 1993, Harper Collins.
** *The Economist,* Re-engineering Reviewed, July 2, 1994.

They reduced the time required to process supplier invoices to a fraction of what it was.

During this period, the manufacturer lost hundreds of millions of dollars. People were not buying their cars and their market share was plummeting. Where was their constraint? It certainly wasn't in their Accounts Payable department. See the previous chapter for a further description of an automotive manufacturer's misdirected efforts on Total Quality.

If your constraint relates to the efficiency of a business process, then reengineering may be the right answer. TOC insists that step one in any improvement process is to identify the constraint.

Core Competencies

This technique looks for those areas (product engineering, people skills, technology, etc.) in which a company or its people excel. In particular, it is helpful to identify the strengths of the company that bring added value and competitive advantage.

In one Theory of Constraints program, the executives of a company reevaluated their strengths in terms of core capabilities. In analyzing all of the problems that they were experiencing in producing custom-engineered, made-to-order products, they realized a shocking but simple truth. Their real strength was in their engineering skills and not in their manufacturing capability.

Once they recognized what their core capabilities were, the Theory of Constraints tools — the Future Reality Tree, Prerequisite and Transition Trees — were very useful in coming up with ideas and detailed plans to take advantage of these competitive skills.

Customer Surveys

Surveys are a great way to find out about your customers' complaints, or UDE's (Undesirable Effects) as they are referred to in the Theory of Constraints. The problem with surveys is that they don't tell you what would happen if you corrected those complaints or filled additional needs. Would your customers buy more product? Would some clients switch from buying competitive products to buying your products?

The Theory of Constraints advocates finding out about customer complaints and needs, and using a Current Reality Tree to perform a deeper analysis. By finding common core problems underlying a large number of customer complaints, you can begin to address problems in a way that competitors do not. This leads the way to both a competitive advantage and more Throughput.

Generally, companies respond to customer complaints directly without eliminating the real underlying cause(s) of the problems. See Chapters 10 to 12 on customer satisfaction for examples and the cure. See the ABC Forge case study for an example of analyzing customer complaints.

Visioning (Mission Statements)

Missions and vision statements point to an ideal — a direction that we want to move towards on a never-ending journey. They are fraught with problems that the Theory of Constraints can help resolve.

Missions and visions are made up of multiple ideals. On specific issues, these ideals often come into conflict with each other or with the procedures and training that people receive. For example, a national pizza chain claims that their mission is "Perfect pizzas, every time". Yet their staff is not trained to verify a customer's order to make sure that the order-taker heard correctly. In order to create a "fun environment for employees to work in" (another part of their mission), they have loud music blasting. While order-takers have fun dancing to the music, they cannot hear soft-spoken clients and often get the orders wrong.

The Theory of Constraints is particularly useful in two aspects of Visioning:

- The conflict resolution technique (Cloud) can be used to resolve conflicts such as those described above.
- The Future Reality Tree can be used to ensure sufficient injections (projects/ideas) to meet the ideals set forth in the vision and mission statements. i.e., it's one thing to have an ideal. It's quite another to have the action plans and projects in place to meet the ideal. Adequate measurements and training are part of every consideration for an effective Visioning approach.

Balanced Scorecard

This measurement technique for managers and executives advocates the use of a broader variety of measures to evaluate management performance. As such, it is saying exactly the same thing as "global optima" says in the Theory of Constraints.

There is an excellent article that describes how to tie local measurements to the broader concepts of Throughput, Net proft and Return on Investment.*

Scenario Planning

Since no one knows what the future will bring, scenario planning suggests looking at various scenarios and preparing managers for what might lie ahead. While we can make some preparations for a variety of different likely events, we can't prepare in detail for all of them. This approach forces managers to look openly at different possibilities and prepare a vision of the future. The benefit to management is that they are typically much better prepared when unexpected events happen.

The Theory of Constraints looks more towards a willed future. Through the Future Reality Tree, it creates ideas (injections) that once implemented, are almost certain to create the desirable effects we want. Contingency plans are formed through the negative branch reservation, which cause additional ideas to be formulated to handle contingencies.

The two techniques, together, are a very effective way of looking at, and formulating concrete ideas, to manage an uncertain future.

Empowerment and Self-Directed Teams

With all of the talk about empowerment, we see more and more executives and managers at companies stressed to their limit. They are working longer hours and are having a lot of difficulty delegating. Furthermore, only 13% of the teams surveyed by a recent Mercer Management study received high ratings for effectiveness.**

"Somehow, we have to get past this idea that all we have to do is join hands and sing *Kum Ba Yah* and say 'We've moved to team work'", says Michael Schrage, author of the book, *No More Teams!*

The goals embodied in empowerment and self-directed teams do not take into account four factors:

---

* "Blending Quality Theories for Continuous Improvement," Harper A. Roehm, Donald Klein, Joseph F. Castellano, *Management Accounting*, February 1995.
** *USA Today*, February 25, 1997, "Why Teams Fail."

1. There is no one left to empower in a downsized organization.
2. There is a host of negative consequences to people and teams who are empowered but not skilled or trained to deal with it.
3. There may be a lot of written and unwritten company policies that are blocking people from true empowerment. Either the policies must be eliminated if they are obsolete or people must be properly trained to understand the intent of the policy and break it as necessary without violating the intent.
4. Measurements must drastically change. Individuals may accomplish great things in spite of the team and teams may accomplish great things in spite of certain team members. This is a negative branch to teamwork that must be resolved in order to gain full commitment of all team members.

The Theory of Constraints provides the means to resolve conflicts over corporate policy and to empower people through the use of the Transition Tree. This tool allows people to really understand the intent of a procedure or policy, and to use some judgment with their new found power. It also provides the Thinking Power to find a good solution for measurements.

However, the issue of not having enough people to empower must be resolved separately. This might require a complete TOC study by itself. The syndrome of not hiring people has become a disease causing a host of other problems inside many organizations.

APPLYING THE THEORY OF CONSTRAINTS TO SECURING THE FUTURE

III

9 Applying the Theory of Constraints to Finding Financial Security

All of the previous chapters laid a foundation in terminology and general thinking. Now let's get down to business. This chapter lays out some of the necessary conditions to achieving long term financial security. Because of interdependencies between direct financial steps, measurements, marketing, employee management and other business functions, this chapter is only the beginning. You will see that taking the right steps, orchestrated by a master plan, *in the correct sequence* has exponential effects on the bottom line.

In a brilliant book on analyzing failure,* Dietrich Dorner talks about the difficulty that many people have in analyzing complex situations. He describes how decision makers cling to false assumptions, even after those assumptions have been proven invalid.

In working with senior executives, my observations confirm those of Dorner. Therefore, before applying the Theory of Constraints methodically to securing the future of our organization, let's examine the assumptions that we're operating under.

The Theory of Constraints contains assumptions about what an organization's strategy should be, and why. In the following several chapters, we will review these assumptions or principles. You will hear from executives who have been successful with the Theory of Constraints, and their views on these assumptions.

---

* *The Logic of Failure*, Dietrich Dorner, 1996 Henry Holt & Company.

Take the time to reflect on your views, relative to these assumptions. In order for you to gain exponential benefits from your efforts with the Theory of Constraints, you either must accept these assumptions or completely and explicitly understand why the assumptions are not valid in your organization. In addition, you will need to develop the new assumptions to replace the old ones. You would be wise to validate them by having them scrutinized carefully by a wide sample of people both inside and outside your organization.

In this chapter, we begin with five principles and assumptions about making money now and in the future — all of the aspects that deal with financial security.

Principle 1

Efforts devoted to making money for the long term must not violate the other necessary conditions for securing the future. These include customer satisfaction, employment security and satisfaction and long term competitive advantage.

If you accept this as a guiding principle, then you will accept the need to reflect on the dependencies between these necessary conditions. For example, short-term, cost-cutting actions that threaten employment security and satisfaction and/or customer value can seriously hurt the long term earnings potential. In this context, layoffs or downsizing for a company that cannot meet next month's payroll might be a tolerable action, although certainly not a strategy for long term secure growth. Layoffs or downsizing to keep a company's return on investment ratio acceptable to shareholders and not absolutely necessary to protect cash flow violates one of the necessary conditions of long term security.

In today's world, we must have employees committed to innovating in ways that explode our competitiveness and in ways that make our customer feel good. In order for employees to behave that way, they must believe that innovations will not result in layoffs or at least, that in the process of innovation, they will learn new skills that will increase their value to the job market. Some executives believe that the employees who do the innovations and the employees who are laid off are two different groups. However, with the elimination of people from all walks of corporate life, this is not the case, and everyone knows it.

Employees of corporations that downsize without impending financial doom all say the same thing. Here is an example: page three of a national business newspaper headlines a daughter who got tossed out of the office with her father on "Take Our Daughters to Work Day" — part of a layoff event in Cincinnati.* The outrage among remaining employees and the general public was well documented in the newspaper stories. Another headline in the same issue talks about an oil company's soaring profits, followed by their announcement of plans to eliminate 4,700 jobs. "It just doesn't make a whole lot of sense," a worker is quoted as saying. "You could see them cutting back personnel and doing different things to get out of a bind. But just to make more profit? I don't understand!"

I remember working with a company that had downsized several years earlier. The trauma was still being felt, by the executives and employees alike. The executives kept a hiring freeze on to ensure that they would never be in that same position again. The company had gone from some 20 vice presidents to 5 vice presidents. Do you think that the workload at that level shrunk by 80%. Hah! The workload actually increased, as the company aggressively tried to get into foreign markets, new businesses and joint ventures with overseas partners.

The results were short-term financial gain and ultimately, financial mediocrity. None of the executives had think time. None of them had anyone left to delegate work to. All of their direct reports were overloaded with 70 to 80 hour work weeks. No one wanted to break the hiring freeze (a company policy disease) because of how it would look to the workers.

> *To carry out the principle is simple. Never downsize or perform layoffs unless you have impending financial disaster. And even then, make sure that you know what your constraint is and that you are not giving up capacity that you will soon need again.*

By itself, neither downsizing nor avoiding downsizing are appropriate as a strategy. For example, a CEO or VP who exercises the principle of "not downsizing" without having an alternative and getting their Board of Directors and the rest of the management team behind it is probably committing professional suicide.

In order for the principle to make sense, you must accept the interdependencies between customer satisfaction, employment security and financial security.

---

* *Canadian Globe and Mail,* May 6, 1995, "Thanks – and Goodbye."

If you keep people on payroll, rather than laying them off or downsizing, there is a serious consideration. It's what you do with the excess labor that's the key to long term financial security. In Japan, for example, in many companies, people who are no longer needed to perform a function are put on special project teams. They are used to either find more efficient ways to operate or to find new products and markets for the company.

The statistics suggest that a focus on Throughput, rather than downsizing, pays off. Fortune Magazine recently pointed out that "Investors' ardor for stripped down payrolls fades quickly. A Mercer Management study of 1,000 of the largest U.S. companies found that the compound annual growth rate of market capitalization for downsizers was about 11% from 1988 to 1994. For the companies that concentrated instead on revenue growth, the figure was 15%."*

If you are concerned about the reactions of the stock market to earnings reports, consider the following. Most investment money today is institutional. All investors, institutional and others, must be educated about your company's approach and why they will get a good return in a reasonable period of time. If you do downsize or lay people off, you require a Negative Branch Reservation (contingency plan) to ensure that remaining employees truly believe that their security is not being violated.

Principle 2

Make Throughput your #1 priority. Make Inventory your #2 priority. Make Operating Expense your bottom priority. Recognize that improvement plans often require that all three are *managed* simultaneously, but with the priority on increasing Throughput.

According to the Theory of Constraints, it is imperative that you move your thinking out of the Cost world and into the Throughput world. There are several logical reasons for this.

For one thing, it *works* as pointed out in the statistics above. For another, you can only cut operating expenses so far. As pointed out in the opening chapter, cost-cutting is worth 1 to 10% improvement. Throughput, on the other hand, can be increased almost infinitely. If you focus on marketing,

---

* *Fortune,* March 3, 1997, "America's Most Admired Companies."

you can also make throughput increases happen quickly and have exponential impact.

Consider this simple example company:

Sales	$100,000,000
Cost of goods sold	$60,000,000
Throughput	$40,000,000

Operating expenses:		
Labor	$10,000,000	
Other	$25,000,000	$35,000,000
Profit		$5,000,000

Now the board of directors comes along and says that we must double the profit next year. That means we need another $5,000,000 in profit.

To get that added $5,000,000, you have two choices — either cut operating expenses or increase sales. Let's consider each one:

- **Cut operating expenses** — Simplistically speaking, you would have to cut operating expenses by $5,000,000. Operating expense has two components — labor and other. Look at the components of non-labor expenses. There are a lot of "fixed" expenses that are difficult to cut. Let's face it. Most executives and managers have already cut more than they should have. Can you lay off a plant? Not very easily. Can you sell any of that used computer equipment that you bought 2 years ago? Sure, for 5 cents on the dollar! That's why most companies looking at reducing operating expense look at reducing labor. In this case, our labor expense is $10,000,000. We'd have to cut our labor force in half to get the desired results. How likely is it that we could do that without impacting our customers, our sales or other aspects of our operations?
- **Increase Sales** — This is our second choice. Many executives do not choose this because they have a skill constraint. They don't know how to do it. In our example company, you could probably increase sales by $8.5 million, a mere 8.5% increase, and not have to increase any operating expenses to get our additional $5,000,000 to the bottom line.

Which do you think is more likely to be successful?
Cutting labor in half or increasing sales by 8.5%?

Making throughput your #1 priority means that much more of the time spent in management meetings will be spent on Throughput. It means that you will allocate more dollars to marketing, in particular to market research and test marketing. It means you will perform an analysis of the markets to come up with immediate action plans and long term competitive advantage plans to increase Throughput. It means that you will focus on ways to get your value recognized in the market, and increase the value provided. A fancy term for this is segmentation. We'll discuss these approaches further in the chapter on Marketing.

The President of a TOC company argues, "Throughput might not be the number one priority if you are already operating at maximum capacity and operating expenses are out of control". If you are operating at maximum capacity, what do you do? The Theory of Constraints says that you must identify the constraint. Why? To increase Throughput, which is contributing to paying all of those operating expenses.

If operating expenses are out of control, what does "out of control" mean? It means we are spending money out of proportion to the Throughput it generates. Perhaps we are even spending money that has no positive impact on Throughput whatsoever. Therefore, the answer is not to blindly cut Operating Expenses, but rather to address the core problem. Perhaps we need to educate people on the relationship between OE and Throughput. Or perhaps we need to change our measurement. Either way, the educational reinforcement is that we spend Operating Expenses for one purpose — to increase Throughput.

For companies that carry physical inventories, we describe focusing on inventory as a second priority. The rationale is that the more inventory you hold, the harder it is to respond to new demands in the market.

Financial security demands that you keep minimal risk of obsolescence, and you are always in a position to respond very quickly to market changes. The less inventory you have in your system, the faster you will be able to deplete it and put new, better products into the market. Therefore, it is imperative that you have a specific plan to keep inventory to a minimum.

For example, in a distribution environment, this requires a distribution strategy that either replenishes frequently enough so that large buffer inventories are not required or that delivers a customized product to the end consumer quickly enough that they are willing to place an order for it rather than expect to buy from stock.

As an example of a new distribution paradigm offered by the Theory of Constraints, the Wall Street Journal describes the General Motors Enterprise project.* This is a Theory of Constraints effort that promises to deliver the Cadillac of your choice to your dealer within 24 hours, 95% of the time. Ultimately, if it works as anticipated, it would cut dealer inventory by up to 80%.

Any car dealer who is aware of the impact of the Internet must either pay attention and be part of the change, or be prepared to lose their business. General Motors is already targeting to cut back 20% of their dealers.

Securing the Future by focusing on Throughput first and Inventory second is a survival issue, and not just a prosperity trip.

There are some notable exceptions to this principle. For example, if you are in a cash flow crunch, your first priority will focus on creating immediate positive cash flow. That probably means that Operating Expenses or Inventory become #1 priority. It still doesn't mean that downsizing is the best solution. Another exception might be at the point of impending competitive announcements. Then, reducing Inventory might become # 1.

Moving from the Cost world to the Throughput world requires a total shift in thinking.

- It means that we accept employee expense as fixed, not variable.
- We look for ways to keep utilizing our work force to generate more Throughput for the company.

Principle 3

Good decisions are based on "global optima." Everyone in the company (but most importantly managers, supervisors and support functions) must be coached and measured so that decisions reflect global optima.

Global optima implies what is good for the overall company (profits, return on investment, economic value added) vs. what is good strictly for one department or function.

I remember a Finance VP telling me that every capital investment they had made the previous year earned a smaller return than if they had just put the

---

* *Wall Street Journal,* February 8, 1995, "GM Expands Its Experiment to Improve Cadillac's Distribution."

money in the bank. What this means is that a lot of projects requiring capital were undertaken to meet local needs, i.e., to make some department or individual more productive without sufficient impact on the company's bottom line.

I see a lot of this kind of behavior with technology. How many companies are investing in the Internet with zero or even negative return? I've seen companies spend millions implementing Electronic Data Interchange (EDI) only to have the wrong information transmitted faster!

I'm not suggesting that automation is not beneficial. I am suggesting that if you do not know your company constraint, you will probably apply automation to the wrong things. Or you will use automation in a way that defies increasing Throughput.

The perfect example relates to some consulting work I did with a Laboratory Health Services provider. Some 30,000 doctors across the country bought their services. The Information Systems department had traditionally provided computer software to these doctors so that they could store and analyze patient test results.

Most of the doctors that I spoke to had several major issues with the software:

- The software did not integrate with their patient management systems. They really wanted to maintain one master patient record. The laboratory system was a separate, independent system. Therefore, if a doctor wanted to see all relevant information about a patient, they had to look in at least two places.
- Most doctors received lab results from multiple sources. Even for a single patient, they might do one test series with one lab, and another series with another lab. This meant that they had to look in several places to get information on one patient.
- Information from different labs was in different formats. If a patient went to one lab for a complete blood analysis, the report would come back with a series of numbers. The same patient, going to another lab for the same test, would also have a series of numbers in the results. The numbers would have no relationship to each other in many cases, because the tests were done on different equipment. This meant that storing the information from different labs in one system was potentially dangerous.

Given these circumstances, what do you think would cause the doctor to use more services from laboratory A vs. laboratory B?

In analyzing this situation, I discovered several factors. However, the survey proved that there would be no increase in Throughput from providing an upgrade of the current laboratory software. While it would handle some minor complaints of some of the doctors, such an upgrade would not add a single dollar to the bottom line of the lab company. Where do you think the lab company's priorities were? You guessed it — getting out an upgrade of the new software.

Financial security requires that every decision to spend money is evaluated by it's impact on Throughput, Inventory/Investment and Operating expenses.* The problem we run into is that people who put together proposals for capital spending or increased operating expenses always have fictitious figures. Of course, we rarely hold them to those figures. Since the vast majority of projects today are over budget and take longer to implement than planned, of course we do not achieve the expected return on investment. And that's assuming that the investment was in the right direction to begin with. So the key question is, how do we prevent this?

> *How do you get the desired effect that most of your people, most of the time, are making decisions that will have a positive impact on your bottom line?*

If you don't have this necessary condition already existing in your organization, then you are looking at implementing a huge paradigm shift in thinking and behavior. Sam Pratt, President of Rockland Manufacturing Company, a TOC company in Pennsylvania, comments,

"Coaching is much easier than measuring [in getting decisions to reflect global optima]. Sometimes, personal optima collide with global optima".

You must identify the core problems blocking this condition from existing. Look for policies, training and measurements as the three pillars of either positive or negative reinforcement of the problems.

Most measurements today reflect the specifics of the job that people are doing, rather than the betterment of the overall organization. Right off the bat, we find several obstacles blocking a good measurement:

- **There is a conflict between global and local optima** — For example, in one real-life situation, an insurance company has a multi-million dollar telemarketing center. The staff generate hundreds of leads each

---

* See the Scarborough Public Utility Commission case study. This not for profit organization has translated the concept of Throughput to Customer Value.

day for the thousands of field agents. On the one hand, the telemarketers are measured on how many leads per day they generate. In order to keep their jobs, they must generate so many leads per day. On the other hand, the telemarketers get a bonus when a lead results in a sale. The immense pressure to generate X leads per day means that 50% of the leads coming out of the telemarketing center are terrible — totally unqualified and ones that are sure to result in 0 sales.

■ **There is a conflict between two global optima** — For example, in this real life situation, the service department of an automotive dealer is measured on keeping parts inventory low. The General Manager is measured on sales and profit, part of which is dependent on repeat customers. Customers do repeat business partly based on their satisfaction with the service department. The fact that repairs are taking longer than last year due to reduced parts inventory has made customers very unhappy. They are going elsewhere when replacing their cars.

■ **Inability to quantify the results at the decision making level** — For example, in a public utility, there are two measurements — adding customer value and keeping operating expenses to a minimum. A utility manager who gets a service call on a weekend must choose between calling in a repair person to perform repairs on the weekend, thus increasing customer value or waiting till Monday. Waiting means paying regular rates rather than overtime rates to the repair person. This increases overall customer value, since expenses are essentially a pass through to the customer base. However, it decreases the specific customer's value.

Measurements, however, are only part of the story of how to enact a paradigm shift to meet financial security. Part II is training . Executives have a totally different mindset than managers, who have a different mindset than supervisors who have a different mindset than employees.

Most executives have an explicit or implicit measurement that reflects global optima. They cannot understand how the people below them can ever make decisions that only reflect local optima in violation of global optima. As indicated by example above, people do that for a reason. Often the reason is a conflict in measurements, or the inability to relate their actions to the negative impact on the bottom line of the company. "Hey, man, I'm just trying to do my job. I don't want to get into trouble," is the attitude of many employees.

In order for employees to connect their actions to global optima, they first must see the global optima as worthwhile. You can imagine the attitude of plant employees in one automatic transmission plant who saw layoffs result every time the plant improved productivity.

The organizations that I've seen who do an excellent job of training employees in this area incorporate several of the following aspects into their "training" strategy:

- Discussion and communication of goals to all levels throughout the company. The goals are communicated through avenues that openly invite two-way streets of communications, e.g., corporate newsletters are a one-way communication vehicle. They can give rise to many questions and to negative reactions. Venting of the negative reactions is necessary to get everyone committed to the goal. Therefore, one-on-one and group meetings and other written and verbal forums allow these reactions to be vented and responded to appropriately.
- Implementation of a method to resolve conflicts between measurements and between different optima.
- Education on how to deal with the negative side effects of any course of action, which may result in financial repercussions.
- Education on Throughput, Inventory, and Operating Expense at all levels. This would include an in depth training program on how every individual's measurements and actions impact T, I, and OE.

Principle 4

Establish a strong visual image of where you need to be. Use Future Reality Trees (or some comprehensive equivalent) for the road map to get there.

In one speech, Dr. Goldratt stated that much long-term strategic planning is about as useful as long-term weather forecasting. The assumption behind this is that many strategic plans and forecasts are wish lists, without the necessary injections to make them happen. Also, according to chaos theory,*

---

* *Chaos,* James Gleick, 1987, Penguin Books.

incomplete assumptions ensure that outcomes other than the ones forecast will actually occur.

However, I can also tell you, from painful personal experience, that the Theory of Constraints leads nowhere without concrete goals in mind. The problem with many so-called strategic plans is twofold:

1. The plan is so full of fantasy figures that the focus becomes making the figures acceptable to shareholders, the board and the executives, rather than the real goal and how to achieve it.

2. The plan is focused on achieving principles or vague, hazy results rather than visual in concrete terms. For example, a goal to make us #1 in service to our customers is doomed to failure. (This was the "precise" goal of a large, national bookseller.) Instead, can you visualize a goal that says:

   - Our customers will never wait more than 30 seconds on the telephone.
   - Our products will have a guarantee that is always 3 times as extensive as our competitors.
   - Every employee in our company will be directly measured in their job performance on customer service.
   - A customer's order will be 100% filled on first request, 98% of the time.

Unfortunately, most corporate planning today lacks the strong visual image of what the process or company will look like when complete. It is crammed into a one or two day session, involving the excursion of executives off site to come up with their strategic plan for the next year, two years, or five years.

As one CEO told me (see Scarborough Public Utilities Commission case study), every year for the past five years, their executive team had been going to a retreat to do their plan. Every year, they determined the list of 75 priorities that needed to be done. Every year, when they looked back on what they had actually accomplished, most of the 75 priorities never got done. So a new list was established and never carried out, year after year.

One useful form of strategic planning to create a much more visual image is to use logic to look at the financial and other impact of different scenarios. For example, one CEO used the negative branch reservation to help determine the future direction of their organization. The company, a producer of auto parts, was being asked by the auto manufacturer to commit huge volumes of production exclusively to that auto manufacturer, and commit significant decreases in price over a period of several years.

The entire management team was outraged at the manufacturer's demands. Their idea was to cut the ties to that manufacturer and go out on their own. The CEO wasn't sure, especially since 60% of that company's business depended on that one auto manufacturer. At stake were several hundred jobs, including his own. Yet he knew that he would have serious repercussions if he forced a decision on his management team that they were unified against.

He took both ideas (a) reject the manufacturer's proposal (i.e., go independent) and (b) accept the manufacturer's proposal, and constructed a Negative Branch Reservation on each one. This technique shows logically the direct effects of ideas. Both ideas had potential for several severe negative consequences.

For example, if we reject the manufacturer's proposal, then we will run out of cash to pay our bills in 45 days. (This was because there was no way to replace 60% of their business overnight — new clients typically took 6 months to 2 years to nurture). If we run out of cash, then we will do massive layoffs. Then morale will be terrible or worse, the union will become a bitter enemy. You can start to get a real picture. We're not dealing with numbers. Rather, we're dealing with tangible effects of our decisions.

Ultimately, there were no positives resulting from the idea to reject the auto manufacturer's proposal. On the other hand, accepting the auto manufacturer's proposal had some serious potential negatives, but many, many positives. For example, the demand for cars fluctuates. If the auto manufacturer experiences a downturn, and can't give the required volume of orders to this parts manufacturer, then the parts manufacturer suffers severe financial consequences. Many of the improvements that they were being forced into by the auto manufacturer in order to reduce their costs were only viable based on a certain volume of orders.

Once the negatives were documented, the CEO was able to present these to the auto manufacturer and negotiate a win–win solution to every single negative.

Principle 5

Lasting financial strength is driven by analysis of the interdependent factors driving an organization, and by the resultant changes which take those interdependencies into account. This means a slower start with a tremendous build up of positive energy.

This describes being on the Throughput curve rather than the Cost curve. The converse is what most organizations do. They look for the quick results, usually derived under pressure, without analyzing the complex system impact. The result? A spurt of positive results followed by a volcanic eruption of negatives.

The principle here is to take the time to analyze all of the factors and interdependencies. Yes, it's painful and yes, it requires the patience of a saint. I recall one executive team that I worked with. They committed 10 days to derive a plan that would align all of the different parts of the organization together. I knew that I was in deep trouble when one of the executives declared that his primary reason for being there was intellectual curiosity.

After 2 days, the executives couldn't stand the amount of time it was taking to go through all of the interdependencies between the functions and processes. They did not do the work, arriving at each session without having thought deeply about their function. They left the legwork to the next layer of management, and decided they would come back at the end of the 10 days and review the outcome. At the end of the 10 days, they were not comfortable with the outcome and did nothing.

The moral of this story is that only a CEO can put and keep the organization on the right curve — the exponential Throughput curve. Managers have too many other things pulling at them. They will easily find an excuse (and a good one, too) to put their minds to work where they want to or where they are comfortable.

Also, without an incredibly ambitious goal to justify the significant effort to learn and apply the five Thinking Processes of TOC, the motivation quickly fades. Alignment of effort between functions is valuable, but only if the alignment is towards a common, mutually respected goal.

Therefore, we need a vision of where the curve is taking us. People will work tirelessly if they feel they are on the track to 25%, 50%, 100% or better improvement.

Summary

There are five principles that reflect the approach that TOC companies take to financial security. If you understand them, and can accept them, then you are ready to consider the next set of principles — those required for customer satisfaction and value. If not, why not take some time to reflect on them? Why did those principles work so well in the companies studied? Why wouldn't they work in your company? What are you afraid of? If you can express your fears as a negative branch, then you will be in a position to take the negatives and eliminate them.

Every set of principles is based on assumptions. Assumptions change over time and also under different circumstances. If these are not the correct principles for your organization, then document what assumptions are different. Surface your assumptions and validate them through extensive scrutiny and dialogue with others in your organization.

In the next chapter, we examine the assumptions behind building customer satisfaction and value. We'll look at what TOC companies are doing, and how their approach differs from traditional approaches.

10 Applying TOC to Building Customer Satisfaction and Value

I n this chapter, I will guide you through some of the underlying TOC principles behind customer satisfaction and the step-by-step approach to build it. We'll look at who we really need to satisfy to build our business exponentially, TOC examples of how analysis leads to different answers than the actions that companies had planned, and the steps leading to customer satisfaction that add significant $ to your bottom line.

We'll start by examining some general principles.

Principle 1

Efforts to increase customer value are only worthwhile if they do not violate one of the other necessary conditions of long-term existence and they result in moving us closer to our goal.

A winner of the Malcolm Baldrige National Quality award several years ago ended up in receivership. Their goal of total quality to satisfy the customer came into conflict with the necessary condition of making money, now and in the future. I see many efforts aimed at customer satisfaction that do absolutely nothing for the company.

We must work to improve our product or service in direct relation to the value it brings to our customers.

If the improvement will not result in more dollars to our bottom line, then are we really adding anything of value to the customer? If the improvement adds value to the customer, then either the market or the customer should be willing to pay for that value.

When Sam Walton, the late Chairman and founder of WalMart, instituted greeters in all of the Walmart stores, it was unique. Customers felt good when they walked into the store and someone said enthusiastically, "Hi, Welcome to WalMart!". The greeters were full-time employees. They were trained to help shoppers, to save them time in finding what they wanted and to answer their questions.

Competitors, not knowing any better, love to copy. The problem is that they forget to copy the important behavior and instead copy superficial aspects. For example, many retailers try to copy WalMart by having a busy cashier shouting a greeting to shoppers as they enter a store. Such shoppers, looking for immediate help, have to struggle to get the cashier's attention. Many of these shoppers have opposite reactions to the WalMart experience. They think about what an insincere place this is to shop, with the annoying forced greeting that offers nothing of value to them.

In another example, a company works to improve its cycle time (loosely defined as the time it takes to move a product through a manufacturing process). This change allows the company to put out rush orders in hours, where it used to take days. This kind of improvement in cycle time is only worth something to the manufacturer if it results in more bottom line dollars. This means either the company changes its policy so that rush orders are sold at a premium, or additional clients are obtained because of the better cycle time, or existing clients buy more or buy more frequently.

Principle 2

You may have a constraint anywhere in the supply chain. Identify the weakest link and improve it.

A steel company may not sell any more steel, without the car manufacturers they sell to improving their sales. A textile manufacturer may not sell any more fabric if their distributors and retailers have the wrong product in the wrong location. i.e., if the end consumer goes to the store and cannot find the size, color or style that suits them, then the textile manufacturer fails to sell more cloth.

This situation is exponentially worse if some other retailer is overstocked on those identical colors, styles or sizes. Now, the manufacturer is losing reorders from the overstocked retailer and losing customer business from the out of stock retailer.

Who is the customer that we're trying to satisfy, anyway? An analysis of the supply chain may be necessary to determine where the customer is, and where the weakest link is. See Chapter 13 on TOC and the Supply chain.

Principle 3

Customer retention requires that we: (1) satisfy customers at least to the extent that they expect; (2) satisfy them better than the competition; and (3) satisfy them better than they expected.

---

1. **Satisfy customers at least to the extent that they expect** — This may require some education for *customers and employees* on what reasonable expectations are. For example, a supermarket chain in New York city surveyed customers and staff on how long they thought it was reasonable to wait in a checkout line before being serviced. Staff thought seven minutes was reasonable. Customers thought 1.5 minutes was reasonable. Since people who live in New York city are under a lot of stress, particularly time stress, this factor of the value of their time became a critical one in winning back market share from specialty stores.

2. **Satisfy them better than the competition** — Substandard products and service will have longevity only as long as there is no competition to speak of.

3. **Satisfy them better than they expected** — The principle here is that, sooner or later, a competitor will challenge your market share. If you want to immunize yourself against the competition, then meaningful satisfaction is a great defensive tool. For example, if I go into a nice-looking restaurant, I expect reasonable quality food and service. If the restaurant owner comes over and inquires about my satisfaction or if I get a free dessert sample, then they have exceeded my expectation. Is this factor sufficient to block a competitor from stealing my business away? Maybe.

Since it is very expensive to acquire a customer, we must ask ourselves if we are taking the correct actions to retain and grow existing customers, vs. spending money to acquire new ones.

Analyzing Customer Needs and Complaints: Finding Root Factors

Many organizations use customer surveys as a basis for judging how to increase customer satisfaction. The problem with many customer surveys is that they provide a lot of feedback on customer complaints, but don't tell you which complaints, if fixed, will result in more business for you. Also, the surveys do not describe the interaction of factors. Improving one factor may have little or no impact. Improving two or three factors may help significantly. Improving a whole lot of factors may mean order of magnitude improvement.

Your immediate customers may not be the end consumer of the product. For example, if you are a distributor, your customers may be the retailers and the end customer is the consumer shopping in the retailer's store. We'll focus here on satisfying your direct customers. We'll look at how to provide customer satisfaction from two perspectives:

1. Satisfying needs or complaints.
2. Recognizing customer individuality — opportunities for segmentation.

Ten Steps to Satisfying Needs or Complaints that Build Profit

STEP 1:   Choose at least 5 current clients and 5 competitor's clients

You need a wide enough sample of clients to make the exercise statistically valid — i.e., to make sure that what you are learning about customer needs or complaints will apply to a broad base of customers. If you are a retailer serving thousands of customers in many geographic locations, the sampling technique must be much broader — multiple locations, more customers in each location, sampling at different hours of the day and days of the week, etc.

Your sample must also include clients of competitors. The requirements to get a competitor's client to do business with you may be quite different than the requirements to get existing clients to do more business with you.

STEP 2: *Document between 5 and 10 common complaints or needs*

You need a wide enough sample of complaints or needs to make the exercise statistically valid — i.e., to make sure that the complaints/needs that you end up addressing apply to a broad enough portion of your business. A good starting point is to survey your sales force and customer service people. You may get some bias here in terms of what sales or service people believe the customer's complaints and needs are versus what the customer will tell you directly. However, to ignore your own employees will prove fatal in the long run.

The more common a complaint or need is, the broader base of customers you will impact by addressing it.

Let's use an example to illustrate how this initial information, combined with a TOC analysis, leads to breakthroughs. Here is a list of UDE's (undesirable effects or complaints (implying real customer needs) of a select group of customers of a children's book manufacturer. These are real.

This children's book manufacturer has dozens of top quality, delightful educational books for children. With topics ranging from mathematics to science to fictional stories, the books are always well liked by parents, children and educators alike.

The manufacturer typically sells through independent consultants who sell through homemaker's book parties. However, for the school marketplace, they used to have a professional field force. Recently, they disbanded the field force and turned the market over to independent consultants, losing millions of dollars in the process.

They have a minute and shrinking share of the marketplace. Here are some of the market's complaints, as voiced by school librarians who are the main buyers in the public school marketplace:

1. The books fall apart after 2 to 3 years.
2. I don't have a representative who calls regularly and whom I can call for service.
3. It takes more time to handle your line of books than some of the other lines.

4. Books don't have the standard catalog coding needed for library placement.
5. Hard cover books are too high-priced for children's book fairs.
6. I have a limited budget ($5 per child per year).

Step 3: *Determine which complaints/needs will likely result in either losing/winning customers or business*

Customers always have complaints or unfulfilled needs. However, many complaints will not result in losing the customer today if they are not addressed and many needs will not result in more business today if addressed.

Assume that anything that would cause a customer to make more money or spend less money has value. In the context of the school librarian example, book fairs make money for the librarian in the form of donations of books to the school library based on sales at book fairs.

To understand which of the many complaints/needs you should be addressing, you may need to perform some cause–effect analysis. For example, suppose a customer tells you that they wish your product were available in more colors. You ask them how much more of your product they would buy if you had it available in more colors. The answer turns out to be "none" since the competitors don't offer these additional colors and customers would be substituting one product for another.

Remember that in this exercise, we're not looking for the "killer" factor(s) to gain long term competitive advantage; we're looking for customer satisfaction factors that will result in moving us closer to our goal.

Any factors that give a customer more time or more money should be taken very seriously. Case Study 2, the Orman Grubb analysis, illustrates this point beautifully.

Once some of these factors are identified, the key question becomes how many customers do they impact.

For example, a large office supply store cashier could not provide $35 change from a $50 dollar bill. The customer had to wait several minutes until they could get change. The cashier said that this happens all the time at the beginning of each cashier's shift. When the manager was questioned about it, he said, "It's not a problem because it doesn't happen very often, and besides, it's a head office policy that each cashier is given minimum cash in their till." When the manager was informed that the cashier said they frequently ran out of cash and had to wait for replenishment, the manager said that the cashier must be mistaken.

One such "symptom" may not lose a customer. Several such symptoms start to cause aggravation. When the symptoms happen repeatedly, you will have a customer actively looking for somewhere else to purchase.

In the book manufacturer example, look again at the list of customer complaints. Which ones do you believe offer a customer more time or more money?

If you guessed that every one of the six factors are relevant, you are correct.

The complaints are superficial symptoms of real problems. For example, when you understand this librarian's complaint, you also understand that competitors show up once every 3 to 6 months, displaying new products and explaining the manufacturer's offerings.

On the other hand, a consultant in this organization is typically a housewife with young children who usually deals in home parties. The consultant typically has no professional sales training and no knowledge of how to approach school librarians and understand their unique needs.

Step 4: Do a cause–effect analysis of what few factors are causing pain

We'll look at two examples here, to illustrate the use of Current Reality Trees to understand underlying problems. Let's look first at the book manufacturer.

Here is the list of complaints. Look at them closely and ask yourself if you can spot any relationship between them. Does any one factor cause (directly or indirectly) any other factor?

1. The books fall apart in 2 to 3 years.
2. I don't have a representative who calls regularly and whom I can call for service.
3. It takes more time to handle your line of books than some of the other lines.
4. Books don't have the standard catalog coding needed for library placement.
5. Hardcover books are too high priced for children's book fairs.
6. I have a limited budget ($5 per child per year).

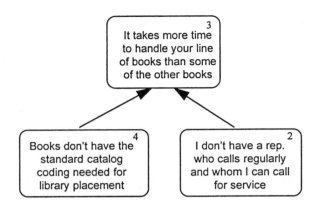

Figure 11   Book Manufactuer's Current Reality Tree

Looking closely at factors 2,3 and 4, we can spot some potential connections. From the librarian's point of view:

> IF books don't have the standard catalog coding needed for library placement, THEN it takes more time to handle your line of books than some of the other lines.
>
> Also, IF I don't have a rep. who calls regularly and whom I can call for service, THEN it takes more time to handle your line of books than some of the other lines.

On understanding the client, you understand the impact that these factors have on the librarian's time. Remember that schools, like other organizations, have consolidated resources, downsized and put more pressure on employees to accomplish more in less time.

There are many manufacturers of children's educational books. The librarians either don't have or won't take the time to study every manufacturer's catalog, track down wandering sales people and make decisions on what books will be suitable for their library.

When books arrive without pre-coded library information, the librarian has to perform all of the extra work to get the books catalogued and physically into the library.

Let's look at some of the other UDE's, and the relationships in a Current Reality Tree.

In this simple analysis, we can probably find one underlying factor that causes all of these UDE's. The underlying factor could be one of the root

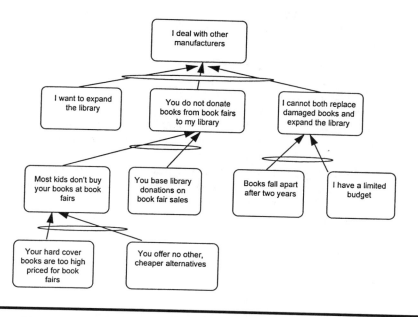

Figure 12   UDE's and Relationships in a Current Reality Tree

problems described in Chapter 5 on the three diseases attacking every organization.

- **Measurements** — No one at this book manufacturer is measured on addressing the school marketplace. This would lead to a chain of cause–effect events which ultimately cause the client's undesirable effects.
- **Training** — No one is trained, or no one within the manufacturer's organization, knows how to address the school marketplace.
- **Policies** — The manufacturer might have a policy, for example, that damaged books are only guaranteed for a certain period of time.

In actuality, this manufacturer has a standard 30-day, 100% satisfaction-guaranteed policy. In addition, you can replace damaged books at half price at any time, without limitation. By extending the guarantee and changing the policy, this manufacturer could easily begin to address the school market. The cost to them would be negligible. The benefits, in terms of increased sales, would be enormous.

Here is another example in a completely different industry. (Note that this entire case study is described in full as Case Study 4 — Strengthening the Marketing Pillars in a Health Services Company):

> A medical lab provides patient testing services to thousands of doctors in a large geographic area. A survey of 1200 of these doctors highlighted that 80% of their complaints related to reporting. For example, some doctors wanted test results grouped by date. Others wanted the results grouped by type of test. Every doctor had their own version of what "stat" meant in terms of reporting. For the very same test, some doctor's wanted to be called at home; others wanted to have a message left with their answering service. Still others wanted an emergency clinic to be notified.

Look at the following current reality tree analyzing the doctor's complaints. This was the first time that any of the doctor's comments had been analyzed rather than just read.

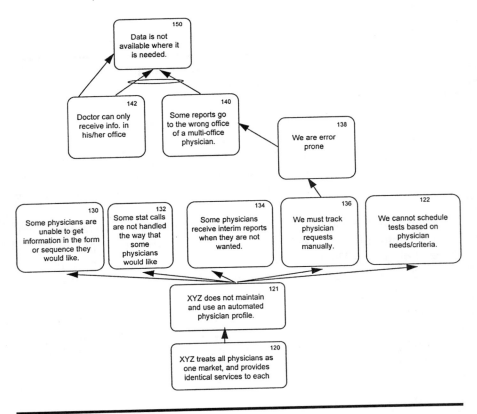

Figure 13 Analysis of the Doctor's Complaints

As we start to make connections, there are a lot of loose ends. I speculated a cause of all of these factors, and ran it by the national management team. They all agreed.

All of these complaints related to one underlying cause. The medical lab, who had invested tens of millions of dollars in computing systems, could only report results in a standard way. The medical director feared being sued if the lab reported results in a non-standard way and those results were misinterpreted by doctors.

The ridiculous aspect of this fear is that it was causing the very problem that the medical director was trying to prevent. Most doctors receive results from more than one laboratory, and no two lab companies report results in the same way. The doctors end up having the lab results manually re-entered into their computerized patient records — a tremendously error-prone procedure.

Furthermore, because different labs use different equipment to perform tests, and because test results are often given in absolute numbers relative to the equipment that performs the test, having lab results from different labs entered into one computer program is a very dangerous practice. A result of "2" on a blood test from one lab might mean "NORMAL." From another lab, a "2" might mean "ABNORMAL."

The systems effort, for less than a million dollars, would have made 80% of the doctor's complaints go away. The equivalent of two more tests per year per doctor served by this lab would have paid for the entire systems effort. Figure 14 shows what the actual current reality tree looked like.

Step 5: Pick a few underlying factors to change

Start with the ones which are either the easiest to implement changes for (example, those that involve changes in your own policies vs. huge engineering changes, or the ones which will eliminate the greatest number of negatives for your clients/market.

In our book manufacturing example, the easiest thing to change (from the point of view of potential speed, i.e., no serious engineering or manufacturing changes required) is company policy.

What would happen if we changed two policies:

1. From now on, books sold through book fairs are 33% off list price. (Or, if it were easy enough, manufacture softcover books for book fairs).

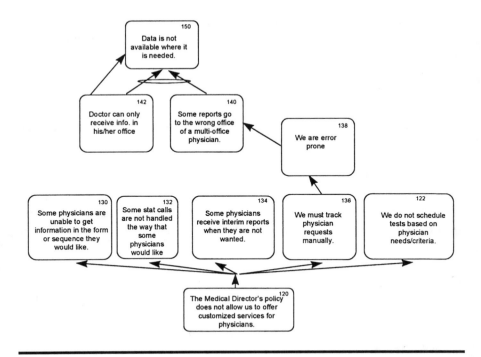

Figure 14   Actual Current Reality Tree

2. Schools have an extended warranty. Their purchases are guaranteed for a full year, and thereafter, they can get any book replaced at 25% of retail.

Would these two policy changes get rid of most UDE's? If our analysis is correct, the answer is yes. Of course, every idea, including the two above, have potential negative side effects that must also be evaluated.

What would be the impact on Throughput, Inventory and Operating Expense?

- **Throughput would go up, probably by a large amount.** However, these two policy changes, by themselves, are not sufficient to cause it. We would need some other injections (ideas) to be implemented. For example, we would require training of the field force on the new policies. We would need a marketing effort towards the schools. We may even want a specialized sales force. All of the required injections would be thought out and vetted in a Future Reality Tree.

- **Inventory would experience more turns, from increased sales and better economies of scale.** Overall, the school environment is probably more predictable than home sales of books. This kind of stability would be a major factor in lowering inventory levels. Also, schools tend to buy at specific times of the year. This is also predictable and can be planned well in advance.
- **Operating Expense would probably increase overall, or would it?** Do we need a new manufacturing plant? Not in this case, where the plant is running only one shift per day and has excess capacity. Do we need another Vice President of Sales? I don't think so. Often, by applying these principles, we can increase Throughput dramatically without significant corresponding increases in operating expenses. That's leverage!

Step 6: *Educate everyone in the organization on the proposed changes and get their input on possible negative side effects. <u>Systematically avoid benefit erosion</u>*

Picture this. A car dealership decides to invest $30,000 per year on loaner cars for customers with major warranty repair work and other satisfaction problems. This is part of a plan to address a host of service problems that lead to customers not returning there to buy cars. A customer is told that they must bring their car in for the forth time to repair the same problem. The customer is upset enough to speak with the general manager of the dealership, who arranges for a free loaner car for several days when the customer brings their car back next week.

The customer's chief complaint is the amount of time he is losing in bringing his car back and picking it up, with the problem not being fixed. When the customer hears that he will get a free loaner, he views this as a benefit. The customer gets up early in the morning to get his car in to the shop. He arrives at 7:45 am and waits in a line of 4 people for his service advisor to service him. The other 3 service advisors have no one waiting, but this dealership has a policy of having a customer only deal with "your" assigned service advisor.

After a 15 minute wait, the service advisor processes the customer and looks for a rental vehicle. Whoops! The only one available hasn't been cleaned yet. No problem. 45 minutes later, the customer is still waiting for the rental car. It took some time to locate it in the lot and then took a long time to get the car cleaned and gassed up.

Result: <u>Benefit erosion</u>. All of the positives that might have resulted from the original benefit of a free loaner car have eroded. The net result is the dealership loses points and the customer is less likely to shop there in future. All of the money spent to provide this benefit to increase customer satisfaction is a complete waste.

STEP 7: *Implement the required internal changes without negative side effects*

In order to determine whether, in fact, addressing a few underlying factors in the current reality trees will have a major positive impact, you must test. Before being able to test, you must simulate the changes inside your company to a few customers.

For example, suppose our children's book manufacturer tried the following test — a policy change on guaranteed replacement to be done at cost (15% of the retail price) rather than half the retail price. Furthermore, in order to test other aspects, they take one of their customer service representatives and make that person responsible for selling to the educational market during the test period. One other change, which the manufacturer determines they can do quickly and inexpensively, is to print soft-cover editions to a handful of their most popular books to have available for book fairs.

Possible negative side effects are identified:

- Customer service workload is impacted, during the testing period, by the loss of the representative. This is particularly negative since the workload was already high.
- People, seeing the cheaper versions of the books, may not want to buy the more expensive versions.

These negative branches must be addressed before unrolling the plan to the sales force or to test customers.

Step 8: *Develop a method of selling these changes to your customers and educating them on what your organization is doing to increase value to them*

Making the changes is not good enough to get significant impact in the market. Customers must be informed in a way that doesn't make your company sound like a politician (i.e., bragging how much you're doing for them).

One method to do this is to help customers understand the analysis of their problems that you've done, and the level of thinking on the solution.

You don't want to leave the impression, with any client, that it's only your company that has these problems. For example, in the medical lab services situation, the doctors whom I interviewed all told me that they had the same problems with all of the labs, not just my client.

Therefore, in a presentation to clients, stress how your problem analysis describes the industry, and not just your company. However, in your description of the solution, this is of course your unique solution to the problem, not the industry's solution.

Step 9: *Execute the sales strategy. Pilot with a few existing close customers*

Now is the chance to find out if your assumptions about the underlying factors causing client UDE's are correct. If they are, you should have immediate sales results. If not, it doesn't necessarily mean that your analysis was incorrect. It may simply mean that you are not marketing it effectively. See Chapter 18 on Marketing.

Any results, other than those anticipated, can be analyzed using cause–effect logic. Why didn't you get the desirable effects? What assumptions were wrong? What entities did you miss?

Once you have a valid action plan, with the expected results that you want and a way to sell them to clients, you are ready for the final step.

Step 10: *Get the buy-in of the sales force to a sales approach that works*

Most customers are resistant to salespeople. Clients are far more educated and sensitive today about the various sales tactics that salespeople use. They have also experienced a host of disappointments, relative to broken promises of companies and unwanted negative side effects of changes. In short, there is often a mistrust of salespeople or a skeptical attitude towards all of the positives that salespeople present.

Therefore, traditional sales approaches with a focus on feature-benefit are likely to meet great resistance. On the other hand, salespeople who are simply told to use a different approach to their customers are also likely to resist trying a different technique.

Neil Rackham's analysis of 35,000 sales calls, presented in his book *SPIN Selling*,* provides some brilliant insights into why many of our assumptions about selling are incorrect. For example, benefits are meaningless unless related specifically to some UDE or problem that a client has and needs to eliminate.

One answer to this dilemma is to have a successful salesperson explain to the sales force why a traditional approach won't work, in terms that salespeople can accept. If this is followed by the results of pilots where a different approach was tried and resulted in success, most salespeople will pay attention.

Remember that many salespeople are also hardened to the tough market, customers that are hard to sell, increasing competition and mass confusion created by more and more competitive product offerings and too many changes inside your company.

Salespeople are often skeptical about corporate management's understanding of real life in the field. One salesperson told me that one of the biggest lies of all time comes from the corporate executive who says, "Hi. I'm from head office, and I'm here to help you."

In another situation, a regional manager of a health insurance company told me that it was often useless to take complaints to head office. "They're corporate people who have grown up in head office. They just don't care what the field thinks."

It saddens me to see the hard work of people fail because one or two of these principles are not taken into account.

Summary

In order for ideas in the realm of customer satisfaction to result in bottom line improvement, they must recognize a multitude of interrelated events and variables. The 10 step process outlined in this chapter provides a structured approach to eliminating many customer complaints and satisfying many customer needs simultaneously. Often, this can be accomplished without major changes to manufacturing processes and without reengineering, simply because it is the organization's policies that often cause many of the customers' problems.

In the next chapter, we'll examine some generic customer satisfaction issues, and some generic approaches to solving them.

---

* *SPIN Selling (Situation, Problem, Implication, Need),* Neil Rackham, 1988, McGraw Hill Inc.

11 Generic Steps to Customer Satisfaction

In this chapter, we will examine some generic steps to achieve "customer satisfaction" that will explode the bottom line. These steps apply across all industries. Using the logic of TOC, we'll illustrate ways to use these steps to make more money, please more customers and make employees happier and more productive, all at the same time.

Three Generic Steps to Customer Satisfaction

Step 1:   Have a full-time customer satisfaction analyst, who is willing to wager half their compensation or more based on finding ways to turn customer dissatisfaction into more profit

The general principal is to train your people and surround yourself with people who know how to make money for the company. That means that your focus changes from "Customer Satisfaction" to (1) customer retention (dissatisfaction and problem prevention) and (2) adding customer value.

I'll share a few examples. In the past year, I was so infuriated with dumb company policies that I wrote letters to the President's of those companies. In most cases, I received a letter or call back from the local manager. In several cases, I received nothing in return. In no case did I receive even a form letter from the President.

The letters I wrote were detailed. They explained why the problem was created by corporate policy and not a local store problem. The comments were obviously ignored. The local managers all admitted that they could not fix the problem permanently, only make amends for the single incident locally.

You are now appointed to be my customer satisfaction analyst. I'll pay you 25% of any additional profit you generate. Here's one letter from an irate customer that will make you $1,000,000 this year, and add $4,000,000 to your company's bottom line. I'll show you the Theory of Constraints approach to handling it.

"Congratulations on the opening of your supercenter in Tampa, Florida. It's a beautiful store with some excellent staff. My wife and I have shopped there several times since it opened.

We do have one problem that has reoccurred every time that we've gone into the store. People do not seem to know how to do basic things.

For example, on our first visit to the store, we bought some items for home delivery and took some items with us. The department responsible for the home delivery items did not know how to handle a home delivery transaction, so they sent us to the cashier. The cashier did not know how to handle home delivery, so she called the head cashier. After waiting 10 minutes for the head cashier to arrive, he indicated that he did not know how to handle the transaction either. The cashier processed the "take-home" goods, and then sent us to another line-up to process the home delivery. It didn't take a brain surgeon to figure out that the procedure for home delivery was quite simple. It was just a matter of knowing which button to press.

On our second visit to the store, we needed to have some carpet cut. We pressed the automatic service button, and heard the public address system announcement for service requested in the carpet cutting area. The announcement went on for 20 minutes, and no one showed up. There were numerous staff walking around the store, but apparently none of them knew how to cut carpet. None of them were concerned enough about the 20-minute loudspeaker announcement to stop by and talk to us while we waited.

On our third visit, we bought an outdoor patio set, and wanted a matching chaise lounge to go with it. The item was not stocked in the store in the color of our set, although other colors were stocked. Does this make sense to you? A person in the department took down all the details and told us he would call the manufacturer and order the product for us and give us a call back. A week later, after not receiving a call back, we returned to the store and were told that this person had quit. Another person took our order again and promised to call us within a day to confirm the order. Again, we never received a call. On each one of these visits, we had to wait 15 to 20 minutes to speak to someone about the problem.

Yesterday, we went back to the store's customer service counter and waited 15 minutes for someone from the department to come to the service counter. A young lady arrived and told us that she couldn't help us because she was just temporarily helping out in that department and didn't know the procedures for "special ordering".

The department manager and another person from the department arrived about 10 minutes later. The other person said, "I've tried to get hold of the Manufacturer for three days and their 800 number is busy." The Department manager told us that if we wanted something special ordered, we would have to pay in advance. However, he couldn't tell us how much we would have to pay or when we would get the item in question.

We asked to speak to a store manager. This was the third time we asked to speak to the store manager and she was not available. Another young man, who was another Department manager, came and talked to us and told us he would take care of our special order. He left a message for us yesterday indicating that he was working on it, but to date, we have heard no other information.

What is the problem? You have people in the store who want to help. Yet we've wasted hours of our time in frustration. Everyone that we've spoken to has tried to put the blame on "high turnover" in the store.

This problem is costing you money. Recruiting is an expensive, time-consuming process. Lost sales is worse. It's great that you reported a 26% increase in sales over last year. But perhaps it would have been 50% without these problems. Also, remember that a good economy can mask your problems. Do you want to be profitable when the downturn comes?

The in-store problems are compounded by the distribution problem exemplified by the manufacturer goods. Also, the practice of having a complete set of patio furniture in one color and a partial set in another color is confusing.

We like your products. Please try to solve these problems so that our shopping experience can be a pleasant one."

We start the Theory of Constraints approach by listing between 5 to 10 customer UDE's (Undesirable effects). In this case, we have some of the store's UDE's that tie closely together with the customer's.

1. It takes too long to get service.
2. It takes too long to get information.
3. People don't meet promises to get information.
4. Staff are not knowledgeable about procedures.
5. Staff turnover is high.
6. I can't get items I need in a set.
7. Your policies make me mad.
8. I have to wait a long time to see a manager.

Unless you have only one retail outlet, we want as broad a list of UDE's as possible. We are looking for the common UDE's to multiple customers, multiple problems and multiple stores.

We analyze and find the core problem — the disease that must be cured (See Figure 15). We'll look for the diseases that have the biggest impact on our bottom line (or other necessary conditions of our business). We'll temper that by looking at the diseases that will give us reasonable benefit within an acceptable timeframe and degree of effort. Another factor is the number of customers that get impacted by the diseases.

As the customer satisfaction analyst, your job is to not just diagnose the problem, but to figure out why it's happening and find a cure. In fact, the cure will have to result in much greater profit, or you don't earn much money!

To understand why so many managers solve problems by firefighting (the Core Problem at the bottom of the diagram), (which, by the way, has been going on for years), we'll use a conflict resolution diagram (Figure 16). This should also give us a starting idea to overcome this problem.

This is a store manager's dilemma. They know that they want to prevent customer problems from getting out of hand (fire damage). In order to do that, they must put fires out quickly. If a customer has a problem, they must deal with it immediately, or it will escalate. In order to deal with it quickly, they must spend time putting the fire out. This is because (assumption) they are the only ones with the authority and/or experience to do so.

On the other hand, to prevent fire damage, they know that they must prevent fires from happening in the first place. This is not so easy. To do this, they must spend time (a lot of time) thinking and training their people. One assumption is that they (the store managers) are the only ones who can do such thinking.

The essence of the conflict is time. With more and more customer problems every day, who has time to do both? No wonder managers deal with problems on a fire-fighting basis!

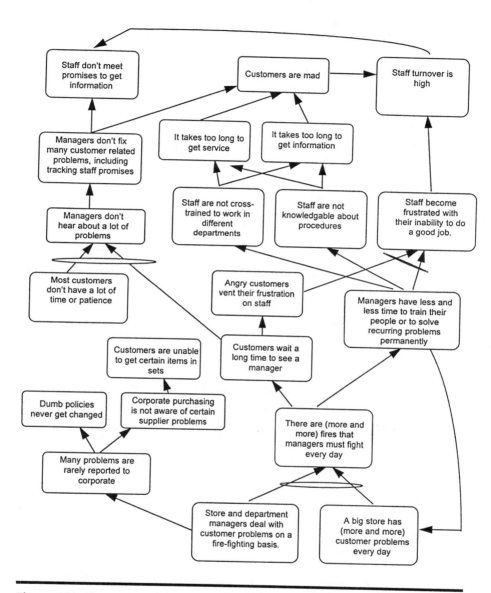

Figure 15   Current Reality Tree of Retail Customer UDEs

Understanding this dilemma and the relationships between the UDE's in the store environment, we can easily find assumptions that can be overcome. For example, the assumption that the store managers are the only ones who can think about how to prevent fires and train their people can be overcome by having this as a dedicated regional or national function.

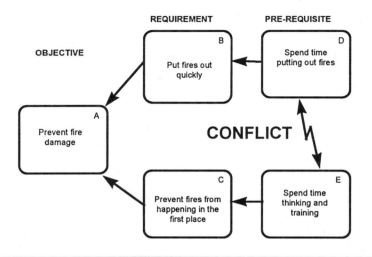

Figure 16   Retail Conflict Resolution Diagram

Provided that we identify how such fire prevention will make money, the function can be justified. So the key is to clearly understand how a customer's behavior and store employee's behavior will change when the problems are fixed permanently.

For example, by how much will employee turnover decrease if we eliminate the identified problems? This could be more accurately predicted by interviewing employees who have left the company. Assuming that less employee turnover reduces recruiting and training costs, might it also have an even bigger impact on sales? Will knowledgeable employees be able to sell more goods and help customers buy more goods?

The full Theory of Constraints approach will take this to its conclusion, with a full, detailed Future Reality Tree and implementation trees. Figure 17 is an example of a Future Reality Tree. The starting idea, from the conflict resolution diagram, is "We have a national fire prevention team."

This Future Reality Tree shows five injections or ideas (in bold rectangular boxes) to change the entire customer satisfaction mentality nationally. The previous list of undesirable effects are changed to desirable effects (shown in dash-outlined boxes). With more discussion with employees and managers, this is the beginning of a $4 million addition to the bottom line. Coincidentally, this particular retail chain is rated one of the worst in its category.*

---

* *Fortune,* March 3, 1997, "America's Most [and Least] Admired Companies."

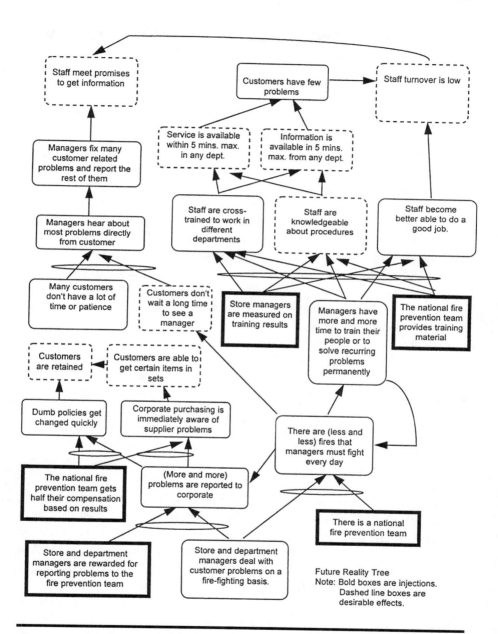

Figure 17   Future Reality Tree

STEP 2:   Provide massive education to anyone who has customer contact on how to do what makes sense.

If you are relying on customer complaint letters to tell you about problems with people who deal with customers, forget it. The vast majority of customers do not complain with their mouths. Instead, they use their feet to go to another supplier, and you never hear about it.

Many people involved in customer contact simply do not know how or are not empowered to do what makes sense for a customer. For example, a business wanted to add another few telephone lines and called their telephone company a month ahead of time. The telephone company representative told the business customer that they would not be able to verify whether or not the business would be able to do so with existing equipment until the day before the scheduled service. If they found out that they couldn't do so, they would have to wait an extra month to order the equipment. "Why can't you give me an answer sooner", the businessman asked. The representative said it was because the telephone company was constantly upgrading area equipment. It was telephone company policy that they only reviewed the customer's request the day before scheduled service.

I'm sure that some people reading the stories here might think that they are ridiculous. However, I'm sad to report that these stories document actual events, and represent a minute fraction of what is actually happening in the real world.

We place too much emphasis on training people on the contents of company policy and holding them accountable to company policies. Most managers forget that the company policy is just a means to an end. It's far more important to train people on the goal of the policy, rather than the policy itself.

The Transition Tree is an excellent tool for training people to think about the meaning of a policy as well as desirable actions to take. It leaves leeway for people to use their brains and their common sense. Figure 18 is an example (a small part) of a Transition Tree used to train customer service people to deal with customer requests that violate policy. See Case Study 1 for another example that deals with company policy relative to employee behavior.

Most people are taught policies as sets of rules and procedures as sets of actions. The important difference with a Transition Tree is the ability to show logic with actions. It shows that every action is in response to fill a need. We expect a result from that action. If we don't get the result, we don't continue. We stop and think. On the right, beside each need and action, is the assumption behind why the action should address the need and get the result.

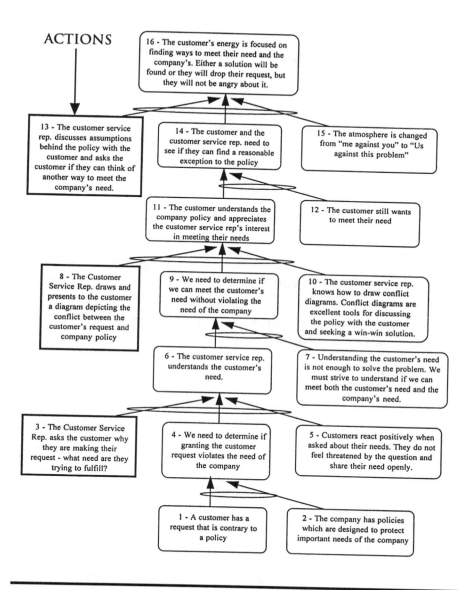

Figure 18 Training Transition Tree Customer Service

We would read the diagram above as follows:

If the company has policies which are designed to protect important needs of the company and a customer has a request that is contrary to company policy, then we need to determine if granting the customer request violates the need of the company.

If we need to determine if granting the customer request violates the need of the company, and the Customer Service Rep. asks why they are making their request — what need are they trying to fulfill, then the customer service rep. understands the customer's need. This is because customers react positively when asked about their needs. They do not feel threatened and share their need openly.

See Chapter 3 to review instructions on how to read a Transition Tree.

The story continues. When such a procedure/policy is explained to a customer service representative, they understand the intent of the policy and customer needs in much more depth. They are not following company gospel blindly.

I have not found a better way to train employees who have customer contact.

Step 3: *Empower anyone who has customer contact to resolve customer issues in the very first contact. Do not pass a customer from one person to another. Do not make the customer explain their story multiple times*

That means retail clerks have the authority to discount or give away items on the spot. It gives airline reservations people the authority to satisfy customers in any way they feel is fair to the customer. It means a contractor can agree to redo a job. It gives an auto dealer service representative the authority to provide a free rental car.

Yes, there are potential negative side effects of empowering people. There are potential abuses. The problem that most operations have is that they have skewed the system towards avoiding possible abuses rather than towards satisfying customers. This is like the judicial system which provides more protection for the criminals than for the victims. Organizations have hundreds of rules to deal with potential abuses and few to satisfy customers in a timely manner.

Often, even when customer service or sales people or managers want to fix problems, they can't because they don't have the authority. When you look at why they don't have the authority, it's usually because the company has a policy which was set up for a good reason that's blocking people from doing their job and satisfying the customer. The assumption that most companies make is that people in the field cannot be given certain authority because it will be frequently abused. Another assumption is that people in the field are not as well trained or versed in a problem as corporate people.

> For example, a Canadian moved to the U.S. He had a credit card from an international retailer and an excellent credit record. His credit card was issued in Canada and therefore was billed in Canadian dollars. When this person purchased goods in the U.S., the purchases were converted to Canadian dollars. When this person paid their bill in U.S. dollars, their U.S. dollars were converted to Canadian dollars and applied to the account. Tired of the constant service charges from converting currency, this individual went to the U.S. retail store and applied to get credit. He gave the Canadian card as a reference. He was turned down because he didn't have a U.S. credit record. This multi-billion dollar international retailer did not have live computer access to their Canadian records and instead gave a blanket rejection. The customer had several other credit cards which would have served as backup credit, but this multi-billion dollar retailer did not empower their credit manager to take such a "risk". Since this customer had just moved into a new home in the U.S. and needed many things, he decided to boycott this retailer. The retailer lost literally thousands of dollars in sales from this one event, and probably millions from similar ones across the country. A letter to the President of the retail chain resulted in a $5 gift certificate from the local store — their attempt at customer satisfaction. The actual effect was customer insult.

In order to visualize how to satisfy customers more than they expect, put yourself in your customer's shoes. Get some unbiased input from your customers, and you will understand your customer's expectations. Don't boast about how wonderful you are until you are sure that your employees can consistently carry out the actions.

Summary

To build more sales and retain customers longer, the focus must switch from customer satisfaction to dissatisfaction and problem prevention and to adding meaningful customer value. To be successful requires the analysis skills to examine combinations of problems, make cause–effect connections and find root causes. Further, the solution must include strategies that will remove the root problem permanently and, in the process, guarantee more Throughput.

In the next chapter, we examine some specific common customer satisfaction problems and TOC approaches to solving those problems.

Some Common Customer Satisfaction Problems and How to Solve Them

I
n this chapter, we examine some specific, common "customer-satisfaction" problems in different industries and the TOC approaches used to solve them. Following are some problems that should NEVER happen, and how to fix them. I encounter them almost everywhere I go. How can managers let things deteriorate to this extent? It must be because these companies are in the Cost world, in a negative situation that is getting worse and worse.

Top Three Customer Satisfaction Problems and How to Eliminate Them Permanently

Generic Problems and Suggested Solutions

1. You get the same customer complaints over and over again.

1. Do a cause–effect analysis to determine the root problem. Look for erroneous rules, skills or measurements as a likely cause. Cure the disease, not the symptoms.

2. Infuriating voice-mail systems that require either:
   - A Ph.D. in voice mail to understand the system.

2. Program the voice-mail system to offer, as the first choice:
   - "Press 1 to complain about this voice-mail system". Have the system then ring into the President's office.

Top Three Customer Satisfaction Problems and
How to Eliminate Them Permanently (continued)

■ You to be retired because it may take you all day to get the choice you want or to speak to a person who can help you.

■ A willingness to leave a message that rarely gets returned.

■ Unlimited funds for holding while on long distance.

■ Have every executive and manager randomly call in, once per month, to use the system, simulating a customer.

3. It takes more than 5 minutes to handle a complaint.

3. Assume that God didn't give bigger complaint-handling brains to the head honcho than He/She gave to many others. There are many people in an organization who have customer contact and could handle the complaint with reasonable intelligence. What's missing?

■ Real empowerment, where authority and responsibility are properly aligned.

■ The definition of explicit procedures and the thinking behind them so that anyone with customer contact can exercise authority with reasonable skill.

■ Enhanced reasoning/logic skills

Retail Problems and Solutions

1. Merchandise is not marked or rings up at the wrong price at the cashier.

1. Give the item to the customer for free. Also, implement a measurement and reward system, as discussed earlier, to cause more people to fix the problem before the customer gets impacted.

2. Cashiers run out of cash; believe it or not, this even happens in banks.

2. Use a buffer management system to ensure that:

■ All cashiers, at any time of the day, have a large enough buffer of cash to handle ANY customer requirement. Use the negative branch reservation technique to remove all significant negative side effects

Top Three Customer Satisfaction Problems and How to Eliminate Them Permanently (continued)

■ Cashiers needing replenishment will AUTOMATICALLY know when to call for the replenishment ahead of when required, and with sufficient notice so that a customer is never kept waiting. One way to make this happen automatically is to have a reserve of cash underneath the drawer. As soon as the cashier goes to the reserve, the head cashier automatically receives a call for replenishment.

■ Master replenishment buffers are within 60 seconds physical walking distance of any cashier.

3. The customer wait time and experience is longer than expected or unpleasant, for checkout, for service, or for anything else.

3. Educate every staff member on the customer perspective and tolerance level, cross-train staff so that people can take over selected functions to alleviate wait times when queue's build, put the manager's office at the *front* of the store, close to where most lines occur, and implement a meaningful measurement system to reward people daily for eliminating/reducing wait times for customers.

Manufacturing Problems and Solutions

1. Late shipments to satisfy customer orders.

1. Implement a constraint-management approach, with related changes to policies and measurements and extensive training to schedule the plant according to what's realistic. Do not rely solely on MRP or Just-in-Time systems to solve this problem.

2. Infuriating Policies — In shipping, ordering, credit, special engineering, etc.

2. Teach everyone with customer contact how to derive win–win solutions to conflicts with customers over company policies. If done socratically (i.e., with a TOC conflict cloud approach), the customer will either come up with a solution that will meet the company's needs or they will voluntarily withdraw their complaint.

Top Three Customer Satisfaction Problems and
How to Eliminate Them Permanently (continued)

3. Defective products or shipments (e.g., short-shipments, wrong products shipped, products do not work on arrival).

3. Compensate the customer beyond replacing the goods (e.g., fine dining certificates, replace with a greater quantity than ordered, etc.). Have a policy and the plant buffer of finished goods to ensure immediate, super-fast replacement of goods. Also, have a measurement such as a pool of dollars or points for a *small* group of people that is severely impacted by this form of complaint (i.e., assume that if you look for quality everywhere, you will have it nowhere. Therefore, the smaller the number of people held accountable, the better).

Distribution Problems and Solutions

1. Delivery times and lead times are too long.

1. Look at:

   ■ The current assumptions that you have regarding costs and methods of transportation vs. cost of inventory. There are likely policies in effect about transportation that may not make any sense in light of current rates of change in consumer demand.
   ■ The current assumptions you have about the numbers and locations of regional and national warehouses vs. the carrying and obsolescence costs of inventory and the cost of lost sales and lost customers.
   ■ How much stock exists on retailers shelves vs. in warehouses. It may be much better to have less stock per store and faster replenishment times, by holding buffers centrally or regionally rather than at every store.

Top Three Customer Satisfaction Problems and How to Eliminate Them Permanently (continued)

2. Out of stock	2. A distributor sometimes forces out of stock situations to occur by pricing goods in step-rate discounts and charging for transportation. What results are fewer, large orders from retailers, some of whom will sit with inventory on their shelves for long periods of time while other retailers have shortages. This hurts everyone in the chain, from the manufacturer who loses sales to the end consumer who can't find an item they want at a local retailer. The way to fix this problem is to buffer inventory differently, through regional warehouses, rather than trying to ship everything in the warehouse to the retailer. This must be combined with changes to measurements and rewards. Also, forget about trying to improve the forecasting systems. For this kind of situation, it is much better to assume that any forecast will have inaccuracies.
3. Unskilled staff relative to basics such as product location, product knowledge, manufacturer returns policy, pricing, custom pricing, status of orders, manufacturer lead times, etc.	3. Training is not sufficient. Incentives in combination with creating a learning organization are more effective. One of the most potent techniques that I've seen is weekly training contests. Each week, someone is selected at random to present a particular product or procedure. No one knows, in advance, who will be selected to present, so everyone must prepare. At the end of the presentation, the person selected is rated by their peers. If they get a passing grade, they are not part of the selection pool for the following presentation. If they get a super grade, they get, e.g., tickets to a sports event or play or dinner for two.

You can add to this list by simply surveying the people in your organization who have ongoing customer contact. The Theory of Constraints approach, if done properly, means that you deal with this situation once and once only.

Summary

Some common customer satisfaction problems will never go away unless we analyze and solve them differently. Some root problems, such as annoying voice-mail systems, demand courageous solutions. Others require the right combination of changes to policies, training and measurements.

In the next chapter, we move into the deeper causes of end-consumer satisfaction, by looking at the entire supply chain.

13 Improving the Supply Chain

I n this chapter, we focus on a methodology for identifying and removing the constraints of your organization, regardless of where they are in the supply chain.

In this case, the chain that we examine consists of the series of organizations involved in a product or service's evolution, from initial raw material to the finished product purchased by the end consumer.

For example, consider automobile manufacturing. Overseas producers and mining companies provide ingots to supply steel companies. Steel and fiberglass producers provide steel and fiberglass materials in various forms to automotive manufacturers, sometimes through several intermediate manufacturers. Hundreds of other manufacturers provide parts and raw materials. The automotive manufacturers combine those raw materials with parts to produce assemblies. The assemblies are integrated and combined with other processes (e.g., paint) and customization (e.g., air conditioning, upholstery) to produce a car. The car is sold to dealers, who may or may not further customize it before the car is purchased by the end consumer.

In any chain, it is the end consumer who economically justifies all of the chain's activities. For example, in the automobile manufacturing example above, if no one buys the cars, everyone from the dealer to the automotive manufacturer to the parts and materials producers suffer.

Every chain has a weak link — something that blocks you from moving closer to your goal. Assuming that the goal is for the entire chain to make more money, and the biggest opportunity to do so is to sell more cars, then let's consider three obstacles that potentially block us, within the framework of a chain:

1. There is not enough supply — someone in the supply chain is choking the supply.
2. There is not enough market demand — we can build more cars. We can't sell them.
3. We have too much supply in some locations and not enough in others — this is the classic distribution problem.

The closer we focus our improvement efforts to addressing the valuable needs of the end consumer, the greater our future security will be.

In order to do that, we must first identify the weakest link in the chain. In some cases, we may not be able to control or improve the weakest link, especially if it is in another organization, and we may have to go around it in order to secure our future.

How do we determine where to focus our efforts in the chain? Here is the step-by-step process, followed by examples:

1. Diagram the chain

If your organization is not at the beginning of the chain, then start at least one level before your organization (i.e., with your suppliers) and continue forward to the end consumer of the product. A detailed picture is not necessary; i.e., if you are a distributor of a variety of products, it's not necessary to show all of your suppliers and all of your customers. Just show in your diagram that there are multiple suppliers, multiple retailers and multiple end consumers. See the example diagram in Chapter 14, which contains an example of applying this methodology. If your organization handles many different kinds of products, start with one product or product group — pick one of the most significant or troublesome.

2. Identify the constraint of the chain as one of the following situations:

- **The constraint is before my organization** — It may be in another division that supplies things to your organization, or in an entirely different organization. In any case, there is no shortage of market demand. Rather, we cannot get enough supply of something to meet the market demand. If we are a manufacturer, we might not be able to get enough raw materials or parts. If we are a service organization, we may be facing a shortage of skilled people. If we are a distributor, we may not be able to get enough product from a manufacturer. If we're a retailer, we're unable to get supply of goods that customers are demanding. *For any of these situations, proceed to step 3A.*

- **The constraint is in my organization** — (there is no shortage of market demand and there is no shortage of supply). We cannot meet market demand. If we are a manufacturer, we cannot produce and ship quickly enough to meet market demand. If we are a service organization, there is some root problem that's blocking us from meeting market demand. It could be the way that we schedule our services, or how we pay our people, or erroneous policies. Whatever it is, we're convinced that there are enough resources available to us, and it is our problem internally to solve. If we are a distributor, a characteristic of our problem is that there is no general shortage of products in the market. Rather, what we have is too much of some goods in some locations and shortages of these very same goods in other locations. If we are a retailer, we may have shelves full of goods that aren't moving quickly enough and other goods that we're not able to keep stocked. We know, from observation and information, that there is no industry shortage of the materials that we have trouble keeping in stock. We have no lack of customers. We just can't service them quickly and efficiently enough. *For any of these situations, proceed to step 3B.*

- **The constraint is in somewhere between me and the market** — (there is no shortage of market demand and there is no shortage of supply). I can get the products out my door as fast as the market needs them, but somehow, the products are not getting to the market in a timely, needed manner. If we are a manufacturer, we are probably facing a problem of distribution or "finishing" of our product — i.e., further manufacturing processes that are causing the problem. If we are a distributor or manufacturer, we may be experiencing some kind of problem with the retail chains. *For any of these situations, proceed to step 3C.*

- **The constraint is in the market** — (our organization has excess capacity to supply the market). We have no overall difficulty meeting market demand, although we may get bogged down from time to time. If we are a manufacturer, we can consistently produce and ship quickly enough to meet market demand. If we are a service organization, we frequently have idle resources. We could handle more clients with no strain on the organization. If we are a distributor, we have excess capacity to supply the market. If we are a retailer, we may have declining or stagnant month to month or annual sales volume. We definitely could service many more customers from our location(s). *For any of these situations, proceed to step 3D.*

3(A). The constraint is before my organization

Therefore, it is probably beyond my direct control and may be beyond my ability to influence. What are our options?

Develop a plan to remove or work around the constraint.

- Can we find additional or replacement suppliers that will remove this constraint?
- Can we work with the supplier to help them remove their own constraint? You can use the Theory of Constraints tools (the Current Reality Tree, Conflict Diagram, Future Reality Tree and Transition Tree) to come up with an attractive offer to your supplier(s) to help them identify and remove their constraints.
- If you have a way that you are convinced will help this organization to remove their constraint (and yours), be very careful how you approach them. Remember, criticism is bad enough. Constructive criticism may turn them off completely.
- One way to spawn their interest is to send a book to some key people in their organization that helps them identify their own situation. Follow this up with an offer to meet with them to help them overcome their constraints. Since their problem is your problem and that of the entire chain following you, you will want to discuss some approaches to remove the constraints.
- Another option is to accept the constraint and find other products that we can produce, reducing our dependency on this supplier for our future security.

3(B). The constraint is in my organization

Develop a plan to remove or work around the constraint.

Hopefully, it is within my control. However, even within my own organization, it may only be within my ability to influence. Our options are:

- Remove this constraint. Go through the five step process (or as many of the steps as are necessary) as described in Dettmer's book, *Goldratt's Theory of Constraints,** or using your preferred process.
- If the constraint is not in your area of control, use your influence to help the function/department in question remove their own constraint. As in step 3A above, if you have a way that you are convinced will help this function to remove their constraint (and yours), be very careful how you approach them. Send some key people a book on the subject, and then follow up with a meeting to discuss some approaches to remove the constraints.
- If you make some effort to convince others to remove the constraint and fail, ask yourself why?
  ⇒ Did they not agree with you about where the constraint is? Why not? Go back and use cause–effect logic to determine whose answer is valid.
  ⇒ Did they agree on what the constraint is, but not on what to change? Why not? It must mean that they did not see how your recommendation would remove the constraint. Go back and use cause–effect logic to determine why they did not accept your recommendation.
  ⇒ Did they accept what needed to be changed and your recommendation, but were not happy enough with the results that you outlined to remove the inertia and get some action going? If this is the case, then you need to look carefully at what results are important to that person or that function. You may need complementary or different recommendations (ideas) in order to achieve results which are important both to you and to them.

Ask for their help to review your logic and show you where you are wrong. By being willing to challenge some of your assumptions regarding your logic, hopefully they will soon be willing to challenge some of their assumptions too.

Remember, it's OK to give people food for thought and let them digest it. Don't give them indigestion by forcing them to eat too quickly. Leave your written materials with them. It probably took you a long time and a lot of thinking and analyzing to come up with your conclusions. Give them the same privilege. Then check back with them a few days later. Keep the discussion open.

---

* *Goldratt's Theory of Constraints,* William H. Dettmer, 1997, ASQC Quality Press.

A note for manufacturers: When there is a constraint in manufacturing, ask the managers, supervisors and shop floor people why. Some typical answers are:

- Machines break down.
- People get to work late or don't do their jobs properly.
- Suppliers don't ship on time.
- We have quality problems.
- Customers changed their minds.

What are these statements? SYMPTOMS!! We can assume that no matter how hard we try to address these, we will NEVER have perfection. Therefore, we should assume that these types of situations will always exist. If we don't want these things to hamper us in the future, we must discover what the underlying diseases are. Using cause–effect analysis and intuition, you may discover that one of the diseases is a skills issue — no one knows how to schedule the plant (i.e., what processes they must buffer) in order to ensure that these problems do not impact us. As a result, there is buffering every-where (quality everywhere, inventory everywhere) and endless expediting.

A note for distributors: When there is a constraint in distribution, ask the managers, supervisors and warehouse people why. Some typical answers are:

- Our forecasting system is lousy.
- The warehouse managers aren't watching their stocks.
- Purchasing bought too much/too little of the product.
- The sales force doesn't give us enough information/notice.
- Consumers behave erratically.

What are these statements? SYMPTOMS!! We can assume that no matter how hard we try to address these, we will NEVER have perfection. Therefore, we should assume that these types of situations will always exist. If we don't want these things to hamper us in the future, we must discover what the underlying diseases are. Using cause–effect analysis and intuition, you may discover that one of the diseases is a measurement issue — Warehouse man-agers are encouraged to keep their stocks low, and therefore try to ship everything out to retailers as quickly as they can.

When shortages occur in some locations and overages in others for the same product, a whole host of negative effects start to happen. As a result, there is unnecessary and deadly buffering at the retail level in terms of how far in advance someone will order goods to replenish low stocks. The farther in advance you order something, the less certain you will be that you are ordering what the market will demand.

Cause–effect analysis will determine the diseases that must be removed and the correct solution and implementation approach.

3(C). The constraint is somewhere between my organization and the market, but not in the market itself

Develop a plan to remove or work around the constraint

Follow the steps in 3A above. Wherever you see the word "supplier", substitute the entity (e.g., manufacturer, distributor, retailer) that describes where the constraint is.

3(D). The constraint is in the market itself

Address the current market constraint

Let's consider two possible scenarios. One is where you already have close to 100% of the market and the market is not growing rapidly, if at all. Also, as part of this scenario, you are unable to add to the existing product/service set that you are currently providing to this market. An example of this might be a public utility providing water and electricity to all residents of a given city. This situation requires you to look for other products/services either in your industry or in another industry. This scenario is not covered here in terms of directing you to a specific type of industry or product set. However, two issues are covered below.

- Once you have found the market that you wish to address, the approaches discussed below will help you figure out how to address the markets more effectively, establish higher customer satisfaction and develop a greater competitive edge.
- In order to meet the necessary condition of employment security and satisfaction, a consideration of getting into new products or markets is to look for opportunities that will take advantage of existing investment in employee skills and capital investment.

The second scenario is where there is opportunity to grab more market share or to add more value to existing clients, in areas related to current products and services. The following approach is used:

⇒ Do a *cause–effect analysis* of what few factors are causing pain to existing customers and to competitor's customers. Concentrate on those factors that will add value to customers and will result in either more orders to you or more profit per order. This analysis identifies a small number of factors (root problems) leading to a much larger number of negatives for clients.

⇒ Segment the market into groups of problems relating to common underlying factors that can be addressed by your organization.

⇒ Quantify, with customer input, the dollar value of additional purchases if the common underlying problems and all connected complaints are eliminated or reduced.

⇒ Pick a few underlying factors to change. Start with the ones that are either the easiest to implement (e.g., ones which involve changes in your own policies vs. huge engineering changes) or the ones which will eliminate the greatest number of negatives for your clients.

⇒ Educate everyone in the organization on the proposed changes and get their input on possible negative side effects.

⇒ Implement the changes without negative side effects.

⇒ Develop a method of selling these changes to your customers and educating your customers on what your organization is doing to increase the value to them.

⇒ Get the buy-in of the sales organization to this method of selling these changes to your customer.

⇒ Execute the sales strategy. Pilot with a few existing, close customers and then expand to other existing and new customers.

See Chapter 19 on Marketing for some recommendations on general marketing approaches.

4. Identify the next constraint, determine if it must be addressed now. If it must be addressed now, do it.

If, by addressing your current constraint, you will end up with excess people resource in your organization, there is a high risk of either pressure to lay off those people and/or enormous job insecurity as people don't have enough work to do to fill the hours available. In this situation, in order to maintain the necessary condition of employment security and satisfaction, we must plan NOW to address not just the current constraint but the next predicted constraint.

For example, assuming that we address our current constraint, what will next block us from improving? Will our constraint shift to the market, or to the distribution channel? This is what we must identify, using our intuition and logic.

5. Go back to step 2.

Summary

Every supply chain has a constraint. That constraint may exist anywhere in the supply chain, from the suppliers of raw material to the primary or secondary manufacturers, distributors or retailers. Depending on where it is, we address it differently. When dealing with the constraints that exist in other organizations (i.e., not our own organization), our approach must be well thought out to ensure that our influence has the desired results.

In the next chapter, we provide an example of applying this methodology to a manufacturer of car stereo systems.

14 Improving the Supply Chain — An Example

A car stereo manufacturer sells directly to the automotive manufacturers as well as to retail automotive chains. They have no supplier shortages that they cannot deal with. They have made several important improvements in their manufacturing processes and quality effort in the past few years. As a result, they have excess capacity.

Excess capacity implies that they must address their market constraint if they want to make substantially more money. Any further effort on internal improvement may unleash some cost improvement. However, in the long run, they will hit a brick wall blocking improvement unless their market constraint is the price of their product and their cost reductions represent a significant part of the selling price. More typically, their internal improvement efforts will ultimately prove futile unless they can break their market constraint.

From consumer interviews, they've determined the following:

With increases in the price of new cars, the end consumer often doesn't buy the stereo of their choice when buying a new car. Buyers of new and used cars are often upgrading after the fact. One thing that blocks these upgrades is the cost of installing components, such as stacked CD players, or better speakers, after the fact. Consumers want plug-in, modular engineering, similar to home stereo components.

The stereo manufacturer approaches the automotive manufacturer to work together to see how the car designs can be changed to allow modularity and easy upgrades to stereo systems in cars. They immediately realize that the inertia and resistance to change of the automotive manufacturers will make it impossible for them to meet customer needs this way (at least in the short term).

Step 1 - Diagram the chain.

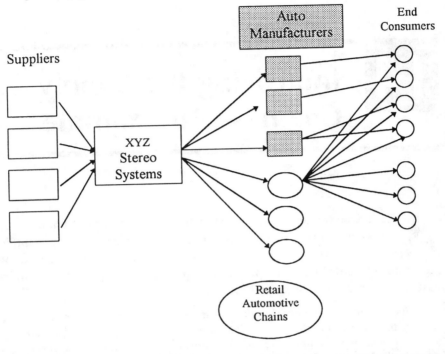

Figure 19   Diagram of the Stereo Systems Manufacturer Chain

Step 1:   Diagram the chain

From this picture, we can see that we have two roads of access to the end consumer — one through the auto manufacturer and one through the retailer.

Step 2:   Identify the constraint of the chain

Clearly, from the above story line, my suppliers and my organization are not blocking me from selling more goods. True, there are almost always cost cutting opportunities with suppliers and internally. The question is, what is the weakest link in the chain? What do I think will generate order of magnitude improvement?

Given a situation where the constraint is after my organization (I have excess capacity), I'd better address that constraint or pretty soon, I'll be under pressure to lay people off. Now we must give some thought to whether the constraint is in the people we are supplying to or in the market.

We have two situations. With the automotive manufacturers, the very design of the car and the placement of wiring to the stereo system makes it difficult to satisfy the end consumer when it comes to upgrading the stereo system. With automotive retailers, we find that the constraint is in the market — the retailers have desire and ability to sell far more of our product — however, the consumer is not demanding it and the retailer is not the one who will generate that pull.

3(C). Develop a plan to remove or work around the constraint.

For the situation with the automotive manufacturer, the approach might be to work together with the automotive manufacturers to change car designs to allow modularity and easy upgrades to stereo systems in cars. The stereo manufacturer knows that automotive manufacturers have tremendous inertia and resistance to change. Therefore, before approaching them, the stereo manufacturer must decide what's in this for the automotive manufacturer and how can they sell these changes to them. Two of the Theory of Constraints tools to accomplish this are the Prerequisite and Transition Trees. These tools are used to look at what obstacles there are to selling such a proposal, and how you might sequence a series of meetings with the automotive manufacturer to gain their buy-in.

For example, it might be relatively easy to convince the automotive marketing and sales VP. It will take dynamite to overcome the resistance in the engineering department. Then you might need an atomic bomb, by comparison, to get manufacturing to go along with it.

All of this implies that you want a work-around, while you execute your master plan with the automotive manufacturers.

3(D). Address the current market constraint.

In this scenario, we want to generate more market demand for our products at the retail level.

⇒ **Do a cause–effect analysis of what few factors are causing pain to existing customers and to competitor's customers.**
   Suppose you find that two of the biggest factors that affect most customers are:

1. They are scared to death of paying more to install a new system than the purchase price of the stereo and
2. They are afraid that, once the new system is installed, if they don't like it, they're stuck with it.

⇒ **Segment the market into groups of problems relating to common underlying factors that can be addressed by your organization.**
The underlying problem may be different for car stereo buffs than for others. In this first analysis, we will assume that segmentation does not add any value.

⇒ **Quantify, with customer input, the dollar value of additional purchases if the common underlying problems and all connected complaints are eliminated or reduced.**
In two days of customer surveys at two different automotive retail chains, we find that we will get 25% more business from the chains if we can address these problems. An upgrade typically has $75 profit in it for us and $75 profit for the retailer. The 25% increase in business is worth $1.5 million to our bottom line.

⇒ **Pick a few underlying factors to change. Start with the ones which are either the easiest to implement changes for (e.g., ones which involve changes in your own policies vs. huge engineering changes) or the ones which will eliminate the greatest number of negatives for your clients.**
We initiate an immediate policy change in order to satisfy the end consumer. We will obtain a list of the new car purchasers with their stereo system, and we will offer free installation of upgrades. The retailer will agree to take a subsidy from us for installation of $15 per unit. This will cover their labor costs, which are sufficient for them given that their sales volumes are also going up and the installation of the equipment does not require a service bay. Further, anyone who installs one of our stereo systems will be entitled to upgrade their system with no additional labor charge if done within 3 months. In this case, the retailers will agree to accept an additional subsidy of $5 for labor since they will get an add-on retail sale, and any upgrade from our existing stereo system will require only about 10 minutes of labor.

⇒ **Educate everyone in the organization on the proposed changes and get their input on possible negative side effects.**
To make such an effort work, we work out the possible problems in advance. Before we turn this over to hundreds of retailers and thousands

of their employees, we want to use our own employee input to make sure that the approach will work without a hitch.

⇒ **Implement the changes without negative side effects.**

A Future Reality Tree analysis is desirable. Also, you must pay close attention to all negative feedback that you receive, and work out ways to overcome significant negatives. It also means piloting the process before rolling it out to the world.

⇒ **Develop a method of selling these changes to your customers and educating your customers on what your organization is doing to increase the value to them.**

In other words, it's not enough to send out letters to customers and negotiate with the head office of a retail chain. We'd better include, as part of our plan, an approach to sell this program to the people on the front lines — the retailer's staff who deal with the end consumer. We must have a line of communication open to these people in case any problems develop.

⇒ **Get the buy-in of the sales organization to this method of selling these changes to your customer.**

Our own sales organization is the one who will approach the retailers and get their buy-in. What salespeople typically do with customers is bombard them with this fantastic new offering we have. The result is typically phenomenal resistance and skepticism.

To get salespeople to sell differently, they must first understand why their typical approach will not work, and then understand a different approach. What is the core problem here? It is a skills problem.

A very different approach that works very well is to present the Current Reality Tree to the buyer, showing how many of their undesirable effects stem from a core problem. You show the buyer why the problem has persisted (the Conflict Diagram) and how your solution overcomes the core problem and all resulting undesirable effects (the Future Reality Tree)

⇒ **Execute the sales strategy. Pilot with a few existing, close customers and then expand to other existing and new customers.**

In the next chapter, we move to another necessary condition of long term security — employment security and satisfaction.

15 Applying TOC to Employment Security and Satisfaction

Every employee in an organization should be contributing something towards the goal of that organization. It sounds simple. Yet when I observe the behavior of many employees in organizations, I see the opposite. It is bad enough to see people not contributing. Many, however, are pulling in the opposite direction.

Here is the underlying assumption behind this chapter.

> *Employees who feel satisfied in what they do and secure in their employment will contribute much, much more to that organization's ongoing well-being.*

Employment security is not job security. It means that individuals have a much higher probability of staying employed in a job of their choice in any organization as a direct result of the commitment of your organization to develop them as individuals.

That doesn't mean that your commitment will do wonders for all employees. It doesn't imply that this is the only necessary condition. It does imply that if you stimulate the *right* efforts to employees who are generally satisfied and feel secure, they will put out a great effort to improve the company.

Let's look at what some of the right efforts mean.

Principles and Assumptions

1. Efforts to increase employment security and satisfaction are only worthwhile if:
   - They do not violate one of the other necessary conditions of long-term existence. These include making money now and in the future, customer satisfaction and long term competitive advantage.
   - They result in moving us closer to our goal.
2. Employee efforts must be tied to a personally meaningful reward that measures their contribution to the organization's goals.
3. The results of the employee's efforts must be visible in stages of days, weeks or months, not years.
4. Layoffs and downsizing which immediately follow or are linked to innovation are the fastest ways to kill any desire an employee has to improve the organization.
5. Layoffs and downsizing in an organization with significant cash (or other) reserves often lead to long term mistrust of management. Often, it breeds mistrust between managers and between managers and executives.
6. Focus employee improvement efforts on Throughput first, then inventory/investment, then operating expenses.
7. Unemployment is bad for our people, our organization and for the society we are a part of. Your employees may not always have assurance of a job with *your* company. It is important that employees feel a strong sense of your organization's commitment to ensuring they remain gainfully employed. How?
   - By developing their deeper understanding of how they can contribute to a stronger, more secure organization.
   - By developing them personally and by developing their job-related skills
   - By displaying, at the top executive level, an understanding that gainful employment and hope for personal economic growth are fabrics of society and business.

Employed people buy products and services and have less criminal tendencies. Employment offers young people hope and an alternative.

Consider this example of an organization which appears not to be focused on overall goals. I talked to a young lady recently who is the best rising star in her organization. I know all the young people in this group, and she stands out head and shoulders above everybody. The organization trained her for 6 months, paying her a salary of several thousand dollars per month, before she was put into the field to start generating business.

In her first couple of weeks, she had booked several thousand dollars worth of business. Since the company does consulting and training work, they were very fortunate that she was able to deliver the training, as well as sell it. There was no one else available to teach when the customer wanted it.

In the weeks and months that followed, she continued to book more in business than the salary she was being paid. As a result, her boss encouraged her to go off her salary and work on straight commission — what he considered to be the next step up in her personal development and career.

Her comment to me was, "There's no way I want to take any risk on this organization. They are totally screwed up. The executives go to regional meetings and fight over pencils — literally!"

Employees who do not respect their management spend a lot of their day looking out for themselves or looking for better jobs. Respect is earned from watching management try to do what's right and what makes sense.

I remember one President I worked for in the 1980s. He had been a great leader while the company was growing. Then, in a period of one year, the company started losing money, customers were complaining more and more, and suppliers were squeezing the company on terms. He began to have serious fights with the two owners of the business.

One week, he spent most of the week going around the company photographing employees. The following week, there was a big billboard in the front lobby with everyone's pictures on it. He had titled it "Employee Appreciation Week." Shortly afterward, he resigned. A year later, this once thriving company was bankrupt.

It wasn't that he was necessarily doing the wrong thing. It was a case of insufficiency and of not doing a mental check on the negative side effects of his actions. The managers who watched him take photos all week likened it to the Captain of the Titanic inviting patrons to sit at the captain's table while the ship went down.

The Theory of Constraints provides some beautiful techniques for changing employee and management behavior for the better. They go along with an underlying belief that most people are basically good.

1. Open The Kimono

The principal is: Everyone should have the right to question things that do not make sense to them. The issue is, how do you do this without creating anarchy?

The Theory of Constraints provides an answer in the Eight Rules of Logic, or Categories of Legitimate Reservations. These are described in the glossary at the end of the case studies. They provide an excellent foundation for questioning anyone's logic.

Opening the kimono means letting other people in on your thinking, and being willing to subject your thinking to their scrutiny. The difference in using the eight rules is that it enables a group to move out of the world of opinion and emotion and into the world of logic. It is not only possible, but probable that you will achieve consensus, if not complete agreement, on any analysis.

The big disadvantage is time, especially as people learn. That's why the Throughput curve leading to order of magnitude improvement gets off to a much slower start. People have to learn how to think!

2. Every policy, every measurement, and every training program is valid for a limited period of time. That period of time is getting shorter and shorter.

Use Current Reality Trees and Conflict Clouds to identify and prevent obsolete or bad policies, training and measurements. Kill them before they kill your organization. Use Conflict Clouds and Future Reality Trees to determine what to replace them with. Use Prerequisite Trees and Transition Trees to implement the changes. Consider these examples:

> One division of a company, in a period of rising sales and profits, changes their entire measurement and compensation system. They create self-directed and empowered teams to service clients, made up of administrative people, sales and systems engineering, service and maintenance people, manufacturing, engineering, etc. 50% of the individual compensation is based on team performance.
>
> In the first year, everyone is handsomely rewarded, as sales continue to rise. The team spirit soars and customers consistently comment positively about the new approach. In the second year, competition and recession in this high-technology segment cause everyone's earnings on the team to take a roller coaster ride downhill. Turnover approaches 50%. Customers don't even bother to complain. They take their business elsewhere. This multi-billion dollar division is put on the auction block.

Look at these policies, which were popular in the recent past. How would your company fare if these policies existed today:

- Just a few years ago, computers were sold with 6 month to 1 year warranties. Today, manufacturers have difficulty selling products with less than a 2 to 3 year warranty, valid worldwide.
- Credit-card applications were long and tedious to fill out, and required several references. Today, many people receive pre-approved credit cards, just by signing a simple form.
- Some retailers are offering to give you items for free if they are priced incorrectly.
- Pizza shops used to have a policy of 30-minute delivery or the order is free. Car accidents and lawsuits changed the policy quickly, although it built huge business overnight.

3. **Good employees value being consulted before being impacted by change. They want to be part of the change process, not victims of it. Consult them *a priori* to save a lot of unnecessary foot surgery from bullet wounds.**

A very effective tool to teach employees for providing feedback is the Negative Branch Reservation. This tool allows them to logically express their concerns.

4. **Guarantee the employee attitudes and behavior you need by careful preplanning and examination of assumptions. Then sequence your communication as a stepladder, to reach the goal at the top. Take baby steps. Bring employees along with your thinking, rather than just telling them the end result of your thinking.**

The Transition Tree is very effective to plan your communications to employees, either in writing or at meetings. Example:

> Company results are not as good as shareholders and the board of directors expect. If the bottom line doesn't improve by 25% within 6 months, you will be forced to chop headcount drastically. Employee morale is low, with salary freezes in the past 18 months. You know that you need a lot of fast innovation in the next few months to have that significant an impact on the bottom line. You are the CEO. Write a communication marking the beginning of a new era for your company. But before you begin, consider what Apple CEO Gil Amelio did to stem the problems.* In a broadcast to employees, as he reviewed the woeful first-quarter numbers and announced the likelihood of more layoffs and cost cutting,

---

* *Fortune*, March 3, 1997, "Apple Computer: Something's Rotten in Cupertino."

he stopped at one point, stared into the camera, and said, "Don't put me in this position again, dammit."

Following are the first few steps of a transition tree that show how planning, assumptions, sequencing and a step ladder of results might flow more smoothly than Mr. Amelio's speech and have a better result:

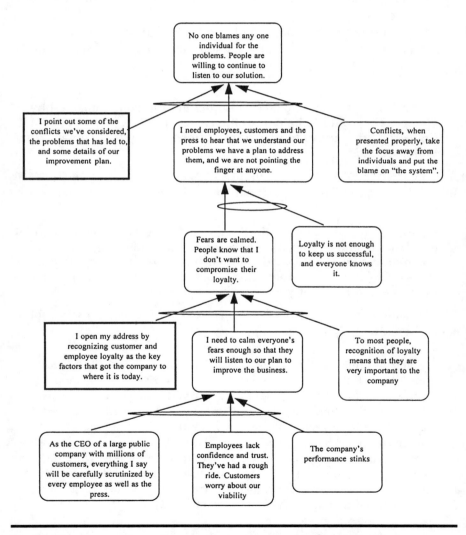

Figure 20   Assumptions, Sequencing and a Stepladder of Results

Reading from the bottom up, this approach suggests that the CEO's number one task is to calm everyone's fears. Why? Because no one will be willing to listen to the CEO's plan if they are running scared. The CEO takes an approach that they believe will calm fears. Why? Their assumption is that "To most people, recognition of loyalty means that they are very important to the company." The CEO won't jeopardize their loyalty.

The next need that the CEO must address is to recognize the problems the company has without pointing fingers at anyone. Why? Because there has been a lot of finger pointing lately and it has led to terrible morale inside the company. Also, finger pointing is not going to solve our problems. To address this need, the CEO uses conflict as a resource to show how concerned the company is about their customers and employees.

Here are examples of conflict resolution diagrams and how they could be used in such a presentation:

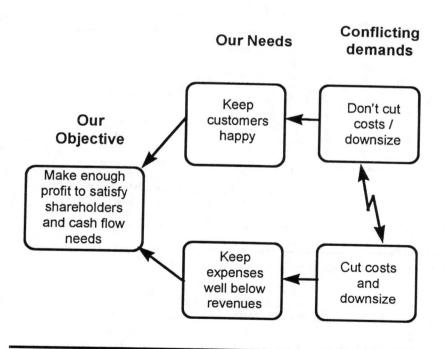

Figure 21   Conflict Resolution Diagram

This is a dilemma that many organizations face. They often make bad assumptions about keeping customers happy and downsizing. By recognizing the conflicts, employees and others begin to understand the problems that an executive team faces in meeting all the needs of the business. They can also look for possible solutions by breaking out of any set of assumptions that this conflict defines.

For example, the conflict states that "In order to keep expenses well below revenues, *we must* cut costs and downsize." If you had the resources and the $1.5 billion cash reserve of Apple, could you think of other ways to keep expenses below revenues? Is it possible that such an organization could increase sales, without having its own products ready to market?

The main point of using the conflict diagram as a presentation tool is to immediately block any finger pointing. It's the circumstances that are to blame, not the individual. This is a non- threatening, educational approach.

Here is another conflict diagram:

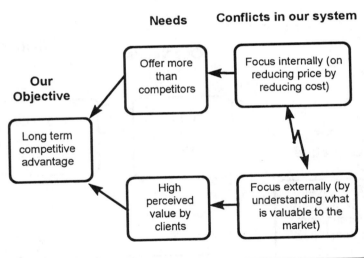

Figure 22   Conflict Resolution Diagram

This conflict exists because the company doesn't have enough resources (or at least the commitment to resources) to focus both internally and exter- nally. Yet the CEO believes that we must meet both needs (to offer more than competitors and to have high perceived value by clients) in order for the company to gain a long term competitive advantage.

Showing employees that these conflicts have been considered and that both sets of needs are important is a major step forward for most executive teams.

Many executives communicate answers, decisions, action plans, etc. They don't consider the process of arriving at those answers important. For many employees, the process is more important than the answer.

Summary

Employment security is very different from job security. By striving towards it with the right efforts, executives will stimulate a much higher level of commitment and much better results from employees at all levels, including their direct reports. Even when organizations take the right steps towards employment security, they must also communicate successfully to all employees.

In the next chapter, we look at how to determine the current reality of your employee attitudes, to help you determine the actions you might want to take.

16 Establishing an Employee Baseline

I n this chapter, I want to help you answer three very critical questions about your employees: (1) What exactly is the behavior of a TOC, Throughput-oriented employee? (2) How closely does the behavior of our employees reflect what we would want a Theory of Constraints, Throughput-oriented employee to do, and (3) How do we cause employee behavior to change?

I demonstrate the answers to questions 1 and 2 in a survey-oriented approach. Here are some questions that an independent survey company could ask your employees confidentially that will establish where your company is today relative to employee satisfaction that counts.

The questions should be asked with answers in four overall categories — Agree, Disagree, Don't Know or Have no Opinion. With each of the two categories where the employee has an opinion, you should subdivide the answers into Strongly, Somewhat, Slightly (Agree, Disagree). Score plus 3 points for strongly agree, 2 points for somewhat agree, and 1 point for slightly agree. Score minus points, similarly, for disagreement.

Some survey organizations would argue in favor of having a mix of questions, where "strongly agree" is not always the desirable answer. The problem that they try to avoid is caused by some people who get into the habit of answering every question the same way, without even thinking. For example, people who dislike the organization might answer "Strongly disagree" to every question. If you or a survey company are concerned about this, you will find that the questions are easy to change. The intent of taking the measurement is still valid.

What's important isn't the starting score, since it is only a number. It is important to target improvement in employee communications and training programs. With the approach as the questions are structured, "improvement" means a higher (towards a more positive) score. A higher negative score means you have a disaster on your hands.

Here is a list of questions, based on my research with TOC companies:

- I know what my organization's constraints are (make sure that you define the word "constraint" so that people know what you are talking about).
- My organization is making progress towards eliminating constraints and improving the business.
- My boss, other executives, and my peers encourage me to question anything that doesn't make sense to me.
- We have a great balance between gaining consensus on what to do and actually getting things done reasonably quickly.
- Organization consistently takes actions with customers that make sense to me.
- Organization consistently takes actions with employees that make sense to me.
- Organization consistently takes action with suppliers that make sense to me.
- Organization is one of the best in the industry or will be within 2 years.
- I can clearly see how my job contributes to organization goals.
- It's fun to work here.
- Management works hard to make sure my ideas are valued.
- I contribute more than I am paid, and the organization appreciates it.
- Customers are extremely important to me. I do everything I can to satisfy them in a way that brings more business in to my organization.
- The organization keeps me posted on what customers expect, and how I can help.
- I get directly rewarded for my ideas that increase company profits.
- Every project that I work on makes a major difference to the company's profits.
- It is rare for our organization to implement something new and have major problems.
- We have techniques that everyone uses to solve problems that don't pit people against each other.
- Everyone is encouraged to talk about the obstacles to making things happen and to help in overcoming obstacles. We actively encourage devil's advocates.
- Everyone in our organization has the responsibility to think.
- Customers must feel satisfied in order for my job and the organization to be secure.
- Any layoffs our organization has done have been fully justified.

- Most days, I feel satisfied with what I accomplish at work.
- My quality of life at work is significantly better than it was 3 years ago.
- My organization will never lay me off unless they have a real crisis.
- If I am no longer employed by my organization, I will be more employable (better trained, better skills, marketable skills) than when I started with the organization.

If most of the employees in your organization strongly agreed with these statements, would you have a better organization, producing more throughput, with more satisfied customers and more collaboration in getting the job done?

These criteria are not intended to be static, nor do I see them as being the only criteria you should use. They are a starting point to measure where you are and to set some strategies for improvement.

What is the best way to change the culture of an organization to a Theory of Constraints, Throughput-oriented culture? How do you change employees behavior en masse?

The best answer I have seen to this question came from Kim Allen, General Manager at Scarborough Public Utilities Commission. He said, "TOC is driven from the top down, with participation into the solutions coming from the bottom up. There are no shortcuts, even though we continue to try."

See the Scarborough Public Utilities Commission case study for more of Kim's advice for executives looking for major cultural changes in their environments:

From my experience, getting people in an organization to think differently takes much longer than most executives expect. Executives go through a transformation, gain some new insights and then typically expect everyone in the organization to follow suit. It just doesn't happen that way.

The first reaction of most employees to major cultural change initiatives is skepticism. This is quickly followed by ridicule (They're crazy. This will never work in our company). Often, there is mistrust (What are the executives up to now? They must have some ulterior motives!). Then comes heavy resistance followed by pretending to accept the new principles.

One example I was witness to involved a customer service organization in a union environment. The customer service workload went exponentially high on the first few days after a monthly billing. Customer phone calls went into the hundreds, with only two or three trained people to take the calls. There were long telephone waits, followed by bitter complaints about the poor service.

When the Vice President responsible for customer service analyzed the problem, he discovered that his core problem was lack of cross-functional skills in this department. This core problem caused a large number of undesirable effects throughout his and other areas of the organization.

This VP had a simple solution — initiate cross-functional training immediately. This would culturally change the whole customer service function. Employees would be more secure, because they would have multiple skills. Customers would be better served because more people would be trained and able to take calls at the beginning of each billing month. The department would no longer be impacted by key people being ill. People who normally answered customer service calls would also be cross-trained to help out other functions in slack time. Throughput would increase dramatically.

All this was true. However, employees who had a confrontational relationship (through the union) with the management had another view. "They want us to do more, be more skilled, without paying us what we're worth. They want us to work at higher skilled jobs at our current base pay rate. Management must have a plan to lay off the more highly paid people. Job security is at stake."

The Theory of Constraints approach to implementing changes impacting employees is a team approach.

First, gain agreement on the goal to be accomplished. There will only be frustration and turmoil if the goal is not agreed to. In order for some people to agree with a goal such as "cross-functional training", we must first answer the question:

"What's in it for me?"

Employees must buy in to the goal. The way that they buy in is by seeing that they will get some benefit out of meeting the goal.

The goals which are the most obvious to managers are often not so obvious to employees. Many people see only the obstacles to reaching the goal, and the negative side effects that the goal might bring. People must believe that both of these issues (obstacles and negative side effects) will be handled to their satisfaction before they can concentrate on a plan to achieve the goal.

Assuming that you have an objective to increase Throughput per employee, and that you want to move employee behavior closer to a TOC approach, you need to undertake a TOC analysis.

1. **Identify the core problems relating to employee skills, measurements and policies or rules. Make sure that you thoroughly understand the links between these core problems and the negative actions and behavior resulting from employees.**

For example, consider the following current Reality Tree relative to employee behavior in an actual organization:

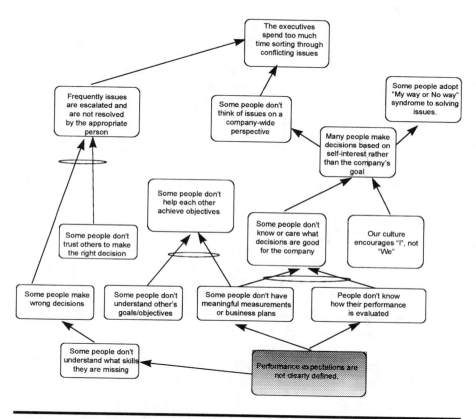

Figure 23   Current Reality Tree of Employee Behavior

In this situation, one of the core employee problems was traced to inappropriate or poorly defined performance expectations. This might really be a training problem (i.e., we don't know how to clearly define performance expectations). This leads to a raft of negative effects, only a small portion of which are shown in this Current Reality Tree.

2. **Determine the conflicts that have prevented the resolution of these core problems. Typically, every core problem that drives employees to do the wrong things has a conflict behind it.**

For example, another core problem in this real-life case study involved the inability to correct the inappropriate behavior of some employees, e.g., excessive overtime, abuse of company property, etc. This also is a skill problem inasmuch as executives had tried all kinds of techniques which did not work. They simply did not know what to do.

Here is the conflict diagram, as they drew it:

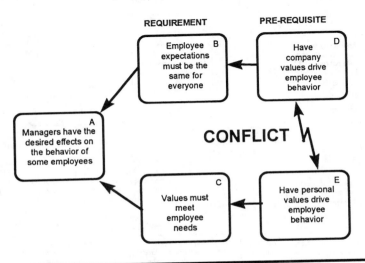

Figure 24 Employee Conflict Diagram

This is a situation common to many companies. Suddenly, a company develops or adopts new visions and changes expectations of employees. The existing employees have probably gotten used to certain ways of doing things, and certain values. As the company's values change, they may hire new employees whose values are consistent with the company's. But what about existing employees?

For example, in one environment, employees received several pairs of new work boots every year, even though they didn't require them so frequently. Employees were given vouchers to use at local stores. They developed a habit of taking every pair they were entitled to, and giving pairs to spouses, friends, etc.

When the company changed their values and put the emphasis on keeping customer costs as low as possible, they tried to get employees to stop taking boots unless they were needed. The company forced employees to prove they needed new boots by bringing their old boots in for inspections. Of course, employees always kept a very old, ragged pair of boots around to prove to the company that they really needed a pair, while they hid the last pair they had gotten from the company. This went on and on, each side trying to outsmart the other. However, the core problem was not resolved.

When the executives drew this conflict diagram, they could immediately see why the issue was never resolved. The problem existed for years because personal values and company values were not linked or aligned in any way.

Many possible solutions come out of analyzing the conflict diagram. For example, you could immediately remove anyone from the organization whose values are not consistent with the company's. There are obvious and numerous negative side effects to that idea.

On deeper analysis, this company found that the conflict existed because some supervisors were not upholding the company's value system — a system which was supported by the vast majority of the employees. The conflict was broken by removing those employees (supervisors) whose values were in conflict with the company's. Some supervisors are just not capable of leading others through a value system that they, themselves do not believe in.

From my experience, the biggest breakthroughs come from understanding the incredible magnitude of problems that result from a single core problem, and the conflicts that have prevented its resolution.

Summary

Before trying to cause changes in employee behavior, it is imperative to first measure the behavior that you are trying to change. Then, set goals. Execute effective programs and measure again. This chapter, in combination with the case studies, provides a measuring tool and examples of how to cause employee behavior to change permanently.

In the next chapter, we examine another of these core problems or diseases that attack organizations and is very closely related to employee behavior — measurements.

17 | Measurements

Many of the problems that I encounter in organizations are directly linked to measurements. 100% of the organizations that I consult with have some bad measurements. The questions that I answer in this chapter are: (1) how to quickly identify (find) the bad measurements in an organization and (2) what measurement system to change to.

How to Identify Bad Measurements

If it is really a bad measurement that is causing a raft of problems, we must be able to prove it through impeccable logic. The most reliable way that I've found to do so is with the first of the five Theory of Constraints Thinking Processes — the Current Reality Tree.

Consider this example. Here are six undesirable effects from a real-life environment — a national health insurance company:

1. Sales are not growing as quickly as we would like.
2. Field agents are frustrated with company-provided leads.
3. Many telemarketing leads have wrong information (name, address or phone number).
4. Turnover of new agents exceeds 85%.
5. New field agents have a hard time earning money.
6. Customer inquiry cards have a low conversion rate (i.e., not very many customer inquiry cards are converted into sales).

The first step in constructing a Current Reality Tree is to connect any two undesirable effects. You can pick any two that have an apparent cause–effect connection. We then "scrutinize" (i.e., apply the Eight Rules of Logic) to the connection.

Here is one connection that is true (Read from the bottom, IF, to the top, THEN):

THEN   Turnover of new agents exceeds 85%

↑

IF   New field agents have a hard time earning money

In this actual situation, the company has, for the most part, excellent field managers and a reasonable training program. Their product is unique and marketable, as proven by the several hundred thousand clients.

So why do new field agents have such a hard time earning money?

New field agents are either very dependent on leads from the company or are generating their own leads, or both. Unless they were in the insurance business before, they don't typically have referrals to launch them into earning a reasonable income.

In this case, the company generates 15 to 20 leads per week per sales agent. They do this from a combination of mail-in inquiries and telemarketing. The mail-ins come from ads in *Entrepreneur* magazine. The telemarketing leads are generated by outbound calls from a national corporate telemarketing center.

A new agent, properly trained and coached, can generate several appointments on their own, without these leads. The company leads are supposed to give the agent a quick start. Why wasn't this happening?

The answer, on investigation, is that telemarketing and inquiry leads are not properly qualified. Why not?

Many of the telemarketing leads passed on to the field were never called by a telemarketer. The ones that were called were not asked the right questions (i.e., questions that would more carefully determine whether they were a real candidate for the company's products). Of course, this fact leads to another negative effect — many telemarketing leads have the wrong information (name, address, etc.). This also leads to field agents being frustrated with leads.

We have another effect. Customer inquiry cards have a low-conversion rate (from inquiry to being sold). This also results from inquiry leads not being properly qualified. The ads that the company runs in *Entrepreneur* magazine imply that anyone can afford their insurance. All that a customer has to do is fill out the card and get a quotation. In fact, this is not true.

Why do we have a situation (ongoing for years!) of such poorly qualified leads?

Once again, an investigation produced some additional facts, which turn out to be causes of the poorly qualified leads.

- Telemarketing is measured on the number of leads produced on time (each week). They also have a small bonus if a lead is converted to a sale.
- The Marketing Director is measured on the number of leads provided to the field each week. He is also measured on business growth.

The insurance business can grow for many reasons. The primary reason for this company's growth is that experienced agents are producing more each year. This is natural, and happens without the help of the marketing department. Experienced agents get referrals and also learn how to sell their product more effectively. Figure 25 is what the current reality tree looks like.

In this industry, big growth requires much better market coverage, as a necessary condition. Even though the company has a few hundred thousand clients, this represents a minuscule share of the market. One way to have much better market coverage is to have many new agents getting off the ground quickly.

The characteristics and results of bad measurements, such as the one shown above are:

- **Micro vs. macro management** — Another way of expressing this is that you have measurements focused on achieving something for the department or function (local) rather than for the company (global). In our example, the focus was on how many leads the telemarketing department could generate rather than on how much conversion there is from lead to business. The attitude of the executive was that "if the salespeople can't convert the leads, we must have ineffective salespeople. Eventually, they will leave the company and new ones will replace them."
- **The measurement does not encourage, or sometimes even discourages, a more effective overall organization** — Cooperation between the field and corporate is lost.

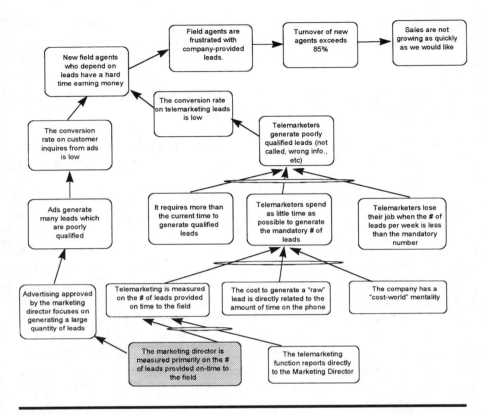

Figure 25   Characteristics and Results of Bad Measurements

- **There is a focus on achieving absolute volume and on reducing cost rather than on effectiveness per dollar spent** — Generate as many leads and inquiries as possible. Spend as little time on the phone per client as possible. This occurs rather than "Generate as many well-qualified leads as possible that convert to business. Spend as much time as necessary to qualify the lead." Sales agents are measured on how many people they saw during the week (how many appointments) rather than how many qualified people they saw during the week.

- **The measurement breeds an attitude of maintaining the status quo and complacency rather than innovation and experimentation** — A great example is the classic measurement of performance to budget. This requirement repeatedly advertised in recruitment ads for managers and executives.

- **The measurement does not encourage CONTINUOUS improvement** — For example, so many of the measurements that I see are based on performance during the year. While results ultimately count the most, the way that the results were attained may determine how *risky* the strategy was and therefore how likely it is that such performance would be repeated. For example, a one-time marketing coup during the year that got the results may not be as valuable as a marketing department that strengthened several areas to achieve results.
- **There is automation of the measurement rather than analysis of the deeper cause–effect relationships** — For example, in quality measurements, there are many cases where the emphasis is strictly on automated reporting. However, the causes of poor quality are much more important than whether Joe is rewarded or punished for putting out 14 good pieces on yesterday's shift.
- **The measurements are mostly internally focused** — There is little or no focus on the client.
- **There is little or no accountability** — For example, many departments are measured on the amount of training they do, and whether or not they meet the quotas and budgets set forth by the company. However, there is no tie in between training and increases in productivity or Throughput.

In contrast, here are some characteristics of good measurements:

- **The measurements drive collaboration between functions to reach a global goal** — For example, in the insurance case study described above, we could measure telemarketing on dollars of business closed rather than number of leads generated.

  The first problem such a measurement might have to overcome is the tendency of a telemarketing manager or a telemarketer to claim that it's not his/her fault that not enough business was closed – the problem is the poor salespeople being hired or the poor training being done in the field, or inadequate selling techniques. i.e., you risk getting a finger pointing behavior.

  In such a case, managers are forced to collaborate, or the problem never gets solved. The question becomes what is the real problem? Is it poor leads or poor field handling of the leads? The answer is not to go to local measurements. The answer is to teach people in the company how to analyze such situations with a stake in finding the correct answers.

The Eight Rules of Logic and the Five Thinking Processes of the Theory of Constraints provide an answer. For example, picture this scenario. The national director of telemarketing and the national director of sales are in a room, trying to figure out how to generate more business with new agents.

The Director of Sales says, "I believe that the reason (cause) why we are having such poor success with new agents is because the leads we get from telemarketing are not properly qualified." The Director of Sales would be required to back up this statement with a number of examples.

The Director of Telemarketing now has a choice. He/she can accept the cause of the problem, and go about resolving it or they can speculate a different cause, or simply use one of the rules of logic (predicted effect) to refute the cause.

For example, the Director of Telemarketing might respond, "If the leads from telemarketing are not properly qualified, you could predict that most agents would not be able to close much business from those leads. But 25% of the leads provided convert to business within three weeks". Note that this is not the situation that I described previously. This is just an example of how a logical conversation, using some ground rules, avoids finger-pointing. The rules of logic change the atmosphere from "you against me" to "me and you against the problem".

- **Measurements focus toward meeting strategic objectives** — For example, in the insurance case study, the strategic objective might be to develop twice as many productive field agents within a certain time period. The measurement could be $ per field agent per month with less than a year's experience.

  With such a measurement, the marketing department is free to address whatever issues are blocking new agents from success (e.g., bad leads, insufficient or poor training, poor time management, etc.)
- **Measurements have a direct, simple impact/relationship with T, I and OE** — particularly profits (T-OE). For example, if we have improvement in the number of productive new field agents, this would directly impact Throughput.
- **Measurements drive *continuous* improvement** — (i.e., in addition to actual improvement in T, I, and OE, they count the **number** of improvements driven from the number of ideas generated).

- **Measurements drive people to think about customer benefits and competitive advantage** — They focus on adding value, from a customer's point of view. Measurements cause people to understand policies, rather than just quote policies to customers. People would be driven to make sure that a policy never hurts a customer.

For example, consider this policy of "Meltdown Airlines", a pseudonym for an airline that is near the bottom of the latest popularity poll. With all of the cutbacks and competition from the airlines, the idea of an "enjoyable airline" may be an oxymoron.

"A ticket issued as an 'L' class ticket cannot be changed without penalty (i.e., charges) to a customer."

Now consider this measurement system:

Tickets which are changed by any field agent without penalty are sent to the Marketing Department of the airline. The marketing department sets the policy, and wants to make sure that any pricing for a seat reflects a segmented market (i.e., a person who pays a reduced price for a ticket should not have the same privileges as someone who pays a full price).

Therefore, whenever the marketing department receives such a ticket, where the customer was not penalized, they contact the clerk who made the ticket change and issue a warning not to do so again. The threat is that another such action would result in a formal reprimand and an entry in the clerk's personnel record. Continued violations could result in dismissal or demotion.

A passenger arrives at the Toronto ticket counter with an "L" class ticket. On this ticket, there are two physical boarding passes representing two flights — one from Toronto to Atlanta, and the second from Atlanta to Fort Walton Beach. He asks the agent to tag his baggage for Atlanta. His wife is meeting him in Atlanta and he will not be using the last leg of his ticket.

The agent says, "I cannot do that. You have a contract to use this ticket as is. If you want to change it, I'll have to reissue the entire ticket at a different price."

The customer explains that he is not changing anything. He is simply not using the last part of the ticket. The Meltdown Airlines clerk reiterates the policy. The customer, getting more upset by the minute, says "Don't bother trying to explain your policy. It will never make any sense to me. If you have to charge me, then do so. I'll just write a letter of complaint to your President."

The clerk responds, "I am making a special note in your file in case marketing comes back to me for not following policy."

The customer says, "I guess customer satisfaction is not part of your policy."

The clerk responds, very sarcastically, "We have to follow policy, but I won't charge you because you are the customer."

The customer, now extremely irritated, says, "So the customer is always right, even when they're wrong."

The clerk says, "You said it, not me!."

What would the clerk's behavior be if the measurement were:

A cash (or some other) bonus for each documented situation where company policy conflicts with meeting customer needs, and the situation is handled by the clerk as a win-win for the company and the customer.

This would of course require some special training for the clerk in order for him/her to be able to use his/her creativity to resolve the problem amicably for both sides.

- **Incentive, rather than punishment, is built into many functions.** The baby boomers were raised on a diet of punishment and respect. Today's generation fears little. Commitment comes out of discovering "what's in it for me."
- **Frequent communications is valued.** CEO's of successful organizations understand how much nurturing it takes to get a message straight and get it through to the organization. As Kim Allen, General Manager of Scarborough Public Utilities Commission, put it, "[you must have] an understanding of the importance of complete alignment, from [the organization's] vision to daily tasks…" This kind of alignment can only come from frequent communications.

With all measurements, it is far more important to work towards improving what you have, than to search forever for the ultimate, perfect measurement system. The TOC methodology proposes finding a "good" solution — one that is practical and easy to implement, and has no serious negative side effects.

Here are some common measurements that I see frequently in TOC-driven organizations:

Financial Productivity

- **T/OE** — The absolute ratio that you calculate from this number is meaningless. It describes, for every dollar that we spend in operating expenses, how much throughput we generate. The objective is to measure where you are and strive for improvement. Note, in this case, that we can improve either by increasing Throughput (the numerator) or decreasing Operating Expense (the denominator). The Theory of Constraints advocates doing both simultaneously, with an emphasis on Throughput first. If you do use this ratio, consider a penalty to the team if the ratio improves but Throughput goes down. This suggests a cost-cutting approach that may also be cutting out the heart of the organization and potential for long-term growth.
- **Throughput** — This is a simple measurement of growth in our products, our markets and also our ability to reduce or contain the direct cost of sales as we grow.
- **Throughput per constraint hour** — Given that we have a physical constraint that is holding us back, every additional ounce of output through a constraint means more sales and more Throughput.
- **Operating Expenses** — The key is to either drive operating expenses down, or, even better, produce and sell more with the same operating expenses. Therefore, an absolute target of driving operating expenses down is nonsense.
- **Inventory/Investment** — The same comments as above apply here. However, there is one additional consideration. Too much inventory blocks us from releasing and selling new products to the market. Too little inventory or inventory in the wrong place at the wrong time also blocks us from selling products to the market. Therefore, any measurement on inventory must reflect this consideration. Further, since inventory is still valued as an asset by banks, shareholders and accountants, any mass reduction in inventory can have a negative impact on the short-term financial reports. This does not imply a difference in strategy relative to inventory. It might imply a timing consideration on how quickly inventories are reduced and the resulting impact on traditional financial reports.
- **Throughput/Investment** — This ratio implies a return on capital. If we include inventory cost as part of the investment figure, it again

provides a two-tiered strategy. One side is to focus on increasing Throughput. The other puts emphasis on reducing or maintaining inventory and investment levels.

Inventory

- **Number of turns** — This measures the marketing and inventory investment strategy effectiveness. Of course, more turns is better, as long as it doesn't come at the expense of increases in operating expense (e.g., transportation costs) which are not offset.
- **Obsolescence** — How much inventory is lost to obsolescence is a good indicator of how well the inventory levels are being managed, relative to market demand.
- **Inventory/Daily Sales (# of days worth of inventory)** — This ratio should be combined with a measurement on stockouts, to determine how effectively inventory is being managed. i.e., it's great to reduce the number of days worth of inventory in the system, provided it does not impact sales. The fallacy in many organizations is that they measure one element to the exclusion of the other, thus fostering local optima. Or they count only warehouse inventory, rather than total inventory in the system. For example, if an auto manufacturer or a computer dealer shoves all of their inventory out to their dealers, and measures their employee only on the amount of inventory in their plants, the measurement causes behavior that hurts the company, i.e., the objective becomes "Sell to the dealer" rather than "Sell through the dealer." When dealers are burdened with high inventories, they either stop buying from the manufacturer or stop paying them.
- **Retail inventory stock outs and surpluses** — For any company involved in a distribution environment, this is a critical measurement of distribution strategy effectiveness. If some stores have too much of a product while others have too little of the same product, the company does not have an effective distribution algorithm. Therefore, we need to measure not just stock outs, but whether we have both stock outs and surpluses of the same items in different locations.
- **Time to replenish** — The longer it takes to replenish goods at the point of sale, the more risk we incur that the lag in response will hurt us. Demand for many items changes so quickly, that our risk of

obsolescence is great if it takes us a long time to replenish. Just imagine the difference between a two-week cycle and a two-month cycle for computer products. In two months, a computer can become totally obsolete.

Customer Value

All of the measurements below are indicative of the value of the customer relationships that we have built, and pointing a direction for improvement.

- **% Increase in the # of customers over last quarter and last year.**
- **% Increase in average sales per order and/or per customer over last year** — Strategically, one of the ways to increase sales is to have customers order more with each order. This may mean greater quantities or more products.
- **% On-time deliveries.**
- **% Stock outs** (For retail/distribution).
- **Demand** — Do we capture, at the point of sale, the fact that a customer asked for something that we could produce but don't have readily available, e.g., a particular size, color, or style combination; for an auto manufacturer, a particular combination of features on a given model car. For a retailer, a particular item that the store does not generally carry.
- **# Complaints per customer and the details** — There are several important factors. First, is the ratio of complaints relative to the size of the customer base growing? Second, are we solving complaints? Do we hear the same complaints over and over again?
- **# Documented cases of customer innovation, relative to complaints** — This really measures whether we are using customer needs, complaints, concerns, etc. as a base to analyze how to gain competitive advantage and respond in an innovative way. Certainly, there is some subjectivity involved in judging whether or not we have innovated. However, if we can document few cases, it points us in a direction for improvement.
- **% Orders delivered in full.**
- **% Returns.**
- **# New products this quarter, this year.**

- **Average frequency of purchase per customer** — Strategically, another way to increase sales is to have customers order more frequently, while maintaining the average value of an order.
- **Lifetime value of a customer** — On average, how long does the average customer stay with our company (e.g., years)? On average, how much does the average customer purchase per year? By multiplying these two factors together, we have the lifetime value of a customer. Not only do we want to increase this figure, as a measure of loyalty, but we also need to understand this figure to determine what kind of investment it's worth to attract new customers. Should we, for example, be paying a much greater incentive to our sales force for cracking new customers?

Operations

- **Average return on capital** — Are we getting better or maintaining our ability to generate profits from our capital investments? This reflects on what projects we approved and competed during the year. It hints at whether or not we've invested in training our people in how to make good investment decisions that impact the bottom line of the business. It reflects on our willingness to take measured risks. It indicates how effective our project management ability is.
- **Average return per project** — This measures whether or not we are learning from both our mistakes and successes.
- **# Constraints broken per chain per year** — This emphasizes continuous improvement, and whether we are striving for several visible steps towards our goal.
- **Collaboration rating by employees** — An employee survey approach measures the perspective of our human capital. Do people perceive a better environment for collaboration?
- **# Throughput projects vs. # of Operating Expense projects vs. # of Inventory/investment projects** — Strategically, what is the organization aiming for? If we want a balance or if we want to put more emphasis on Throughput, for example, then a measurement such as this tells us whether or not our strategic objective is being acheived.
- **Throughput/Project Capital invested in infrastructure** — This measures the strategic value of the infrastructure investment. For example, Edvinsson & Malone describe in their book, *Intellectual Capital,*

the importance of infrastructure to the future success of an organization. They state that this "includes such factors as the quality and reach of information technology systems, company images, proprietary databases, organizational concepts and documentation".* On the other hand, I've seen a lot of money thrown in the garbage can on building infrastructure that had no discernible value to the organization. This measurement points out whether or not we know what we're doing when we make infrastructure investments. We shouldn't have to wait years to reap the benefits. Saying that the benefits will take years is another way of hiding behind a terrible investment.

The Acme Manufacturing case study provides more details on a Throughput-oriented measurement, reward and reporting system.

Summary

The measurements of your organization must be rethought — from top to bottom. The development of new measurements must allow for everyone's input on and thinking through the negative side effects of any new measurement. New measurements which have the characteristics described above have had dramatic effects on organizations. For example, Acme Manufacturing reported an increased market share of 17 to 25% in various divisions, sales increases of 23%, and reduction of customer complaints by 47% as a direct result of overall strategies that involved significant changes in measurements.

In the next chapter, I review why companies (even TOC success stories) make terrible assumptions regarding how to market. I also suggest how to combine TOC and other techniques for a much better answer to the marketing conundrum.

---

* *Intellectual Capital*, Leif Edvinsson and Michael S. Malone, Harper Business, 1997.

18 Marketing

You can make many mistakes, operate with many false assumptions and still keep an organization going. However, in marketing, bad assumptions can literally kill the company. Unfortunately, bad marketing assumptions are a fact of life in many TOC companies that I interact with. Therefore, I conclude that simply using the Theory of Constraints does not correct this problem. I believe it is critical to understand why not, and then to formulate a plan with a much higher probability of working.

I have spent half of my professional life on sales and marketing issues, right through a senior executive level with responsibility for several hundred people. In January, 1997, I attended a marketing seminar by Jay Abraham. The seminar forced me to challenge every marketing belief under which I operated for many years. Based on empirical evidence, it appears that several of my assumptions were either wrong or certainly not valid for many situations.

Consider these examples:

- **How would you coach someone to write a direct sales letter to your clients?** — The kind of letter that most executives write looks nothing like the kind of letter that Abraham would write. Most executives have a primary concern for image. When communicating with customers, the tone of a letter is often "Here's what we're doing for you." Abraham's letters are exciting to read and are all about the results that the client can expect, with detailed, technical back up as to why those results will be achieved. The Theory of Constraints Prerequisite and Transition Trees are perfect tools for writing such letters. For example, the Prerequisite Tree identifies all of the obstacles that you need to overcome in the buyer's mind in the letter. It tells you what condition you must achieve before you write anything further. And it tells you the correct sequence in which to overcome the obstacles. The Transition Tree makes it easy to take the stepping stones from the Prerequisite Tree as results that you need to achieve in the letter and turn those into a series of paragraphs to reach each stepping stone.

Another example of an assumption that must be challenged relates to whether a letter should be brief or detailed. According to Abraham, direct-sales letters have much greater success when they are packed with detail. This often means the "letter" is 5 to 20 pages long. I had always been taught that people were very busy and busy people would not read more than a page or two. While the assumption that "people are very busy" is correct, the second half of the assumption — "that busy people would not read more than a page or two" is incorrect.

In fact, busy people decide in a matter of seconds whether or not to read something. If the headline or first paragraph sounds of direct interest to them, they will read great detail on the subject. At the least, they will browse a long letter. At best, they will devour it because someone finally is providing them with enough information to understand not just the benefits of a product or service, but why they are almost certain to attain those benefits.

- **Out of sight — out of mind.** How often do we communicate directly with our clients? Of course, if they have ordered from us in the last month, we're in touch. What about those that haven't ordered? According to Abraham, we should consider communicating once per month. Consider a national Pizza chain. They have a mission to increase sales based on a new product they are announcing. They are spending tens of millions of dollars in advertising. Yet they have no plans to use their biggest asset — their customer list. In one local store, the customer list consists of 12,000 names. Only 1,500 of those have ordered from this store in the last month. Abraham suggested that one of the most certain ways to immediately increase sales was to go back to your customer list and start phoning or writing. I was skeptical, but I took my customer list and went back to clients I had not done business with for 2 years. I wrote letters and immediately generated business and inquiries.

- **New customers are worth a much higher incentive to the sales force.** What is the lifetime value of a new customer? We often pay salespeople incentives for attracting new clients. The incentives usually bear no relationship to what a new client is worth to the company. For example, if the average customer is with us for 5 years, and buys an average of $20,000 per year, that's $100,000 in sales volume for every new customer. If our margin, after direct cost is 35%, then every new customer is worth, on average, $35,000 in Throughput (or $7,000 per year).

An incentive, for example, of $1,750 to a sales person for a new client would be paid back in profits 4 times over in just one year. Yet, what I often find are meaningless bonuses (i.e., dollar bonuses that don't reflect the effort and risk involved) to salespeople for attracting new clients.

- **Do you have a unique benefit proposal?** Can you express, in one or two sentences, what you *uniquely* (compared to the competition) accomplish for clients? Is the proposal *meaningful* to your clients? For example, an ice cream manufacturer in Texas uses the phrase, "We sell what we don't eat" to describe how unique and good their ice cream is. Another example is a seminar sponsor who promises that "This seminar will teach you techniques that will double your production within one year, guaranteed".

- **Risk reversal.** To what extent are you undertaking the risk of purchasing/using your products & services, rather than having your customer take the risk? For example, on consumer products, guarantees are a fact of life. However, there is a huge difference between a one year and a three year guarantee on a computer system, for example. There is a huge difference between a hardware manufacturer that guarantees the computer hardware and one who guarantees that a whole range of software will also work to specifications on that hardware. There is a huge difference between a guarantee to repair a car, a guarantee to replace a car and a guarantee to take back a new or used vehicle within 3 months and refund most or all of the purchase price.

While these are just examples of assumptions that we must surface, I learned a great deal from the fact that of the 250 people attending this session, very few had ever challenged their sales and marketing approaches. For years, they had been doing things the same way, using the same assumptions and beliefs.

From a Theory of Constraints perspective, there are two important points:

1. Sales and Marketing assumptions by which we operate our organization must be surfaced and examined for validity.

2. Experimentation is often the only way of validating which new assumptions will work. In other words, rules of logic may fail to help when coming up with ideas because the human behavior of a large set of customers is hard to predict. Therefore, we must experiment at a low cost to try out different approaches and validate the ideas that we wish to use in our Future Reality Tree strategies. The issue is how to experiment cost effectively.

Often, sales and marketing assumptions are held and used by one person (e.g., the top marketing executive) or by a small group of people. In marketing, it's very easy to hide behind assumptions based on "experience." "I'm the expert — I've made all the marketing mistakes there are to be made. No one can tell me how to market better."

The problem is that market dynamics change so quickly and drastically, we have no way of knowing how valid or long-lasting our assumptions are. Also, there are so many different ways of marketing, it would be unusual not to have alternatives that would also work or may even work better.

Let's assume that your goal is to generate more throughput by removing the market constraint. In order to find your constraints in sales and marketing, try breaking the goal down into four parts — i.e., the four opportunities to sell more, which include:

1. Increase the number of customers for existing products, either in existing geographic areas or by expanding geographic areas.
2. Increase the volume of sales per transaction, either by offering new products that go hand in hand with existing products, or getting customers to buy more of the same item.
3. Increase the frequency with which existing customers buy (# of orders per year). This technique tends to work better for a service than for hard goods.
4. Seek new markets — new customers for new products.

Do you have constraints in all four areas? Is your biggest constraint

- That you don't have enough customers?
- That they don't buy enough each time they order?
- That they don't order frequently enough?
- That your existing product sales are peaking or dying?

If they don't buy enough with each order, do you have policy constraints blocking volume purchases or are there penalties for the customer (e.g., greater risk of product obsolescence)? Or is the problem related to not having enough products to sell to that customer?

Each of these avenues is an opportunity for a current reality tree study of the market, and of existing customer's needs, as described in Chapters 10 and 14, and in the case studies. Is the total market blocked from buying more of the product from any of the existing suppliers, or are they just blocked from buying more from you?

The implication of what to do about a constraint differs, depending on where the constraint is. In the first case, we have an implication that the market for the product itself is stagnant or declining. This may imply taking action to find a replacement product or to put emphasis on acquiring competitors or their market share. In the second case, the implication is that the market itself is growing. You need to find out how to capture that growth to get your fair share or better. Your marketing approach may be the constraint.

Following are some of the traps, and how to avoid bad assumptions and a host of TOC mistakes.

1. Are we selling products or benefits?

Engineers, especially, beware! Countless TOC expert companies put out some of the worst sales material and presentations I've ever seen. The assumption is that we have a superior product — everyone should recognize it. Today's fact of life is that even today's best products will be under siege from the competition tomorrow. The correct assumption is that we are not selling products or services. We are selling the results that those products or services provide. Therefore, for example, a brochure advertising "Supervisory skills" does nothing to attract buyers nor distinguish us from the competition. A brochure promising "supervisors who make factories produce greater quantities of quality goods without labor strife" is much more specific and tangible.

2. What do we need to produce and sell to be (more & more) successful?

The constraint may not be related to what we currently sell, but to the fact that we must add new products and replace dying ones in order to improve. One very bad assumption that many companies make is that "we must produce the product(s) we sell." There may be other organizations out there who have products that they would love to sell to your clients, but they can't afford the marketing effort. Whatever the problem related to the question above, the answer can be found using a Current Reality Tree analysis of the market and it's needs (being sure to include both your clients and those of competitors).

3. Segmentation

A market is effectively segmented when you can sell exactly the same product at two different prices to two different markets without having either market impacted by the other. The perfect example is airline seats. The same seat can sell for a whole variety of prices depending on how far in advance it is booked, whether or not you are flying over a Saturday night, what privileges you want to change or cancel the seat, etc.

There are some very bad assumptions made about segmentation. One of these is that markets can be segmented without negative effects. For example, a contact lens manufacturer thought that they could sell exactly the same lens for different prices, depending on what they called it. They offered three different packages — one package said the lens was good for a week and was very inexpensive. The second package guaranteed the lens for a month and was moderately priced. The third package claimed the lens would work for several months and was very expensive.

When the word was leaked to the public and the press that these were all, in fact, the very same lens — the only difference was the price and packaging — the product was ostracized and the company's reputation suffered greatly.

Segmentation can be a fantastic strategy to boost the amount of dollars received for the same product by differentiating based on some other factor. For example, a computer chip manufacturer can get several times the price for the same chip based on delivery requirement. One day vs. three week delivery makes a big difference in pricing.

The key to any segmentation effort is to first identify the factors that provide opportunities for segmentation for your products or services (e.g., different lead times for delivery, pricing based on quantity ordered, different levels of willingness to commit orders or quantities ordered). The segmentation is based on identifying the different needs of different markets for the same product.

Then, open up the idea of proposed segmentation to the most negative people inside the company. You want these people to raise negative branch reservations, allowing all sources within the company to provide input on possible negative outcomes. Second, you must do test marketing to test the assumptions before rolling out the program in full.

4. Image advertising is worthwhile

Advertising agencies perpetuate all kinds of myths about image advertising. They love image advertising because a client can never measure the value of an ad in terms of increased sales. There are many image ads which won awards for the advertising agencies yet failed to yield any tangible results for the companies.

5. No one wants this product/service

Take the same product or service and write two different headlines about it. Or try different prices. When you market the product using the two headlines or prices, you may find one has 5 clients out of a hundred responding positively and the other has 5 clients out of a thousand responding positively. I can guarantee that with today's knowledge of psychology, you will be hard pressed to ever understand why this happens. There is no apparent logic to it, although given enough time, research and the TOC tools, you might, eventually understand it. The point is, unless you do at least a half dozen or so tests in marketing every product or service, you will never discover the approach that works well out of all of these. You may conclude, incorrectly, that there is simply not enough interest to market a new product. You may also conclude, incorrectly, that you should abandon an existing product for lack of interest.

6. Coverage

What is the market that you want to reach? What is blocking you from reaching that market? I recall an assignment with Cisco Systems, one of the world's largest internetworking companies. The General Manager wanted coverage on his market. He defined it as the Fortune 500 plus the top 20 organizations in every vertical — utilities, manufacturers, banks, insurance companies, etc. Within a few days, the target organizations were identified. Within two months, the top three people in each organization were identified. A telemarketing script was developed and a campaign was launched. Within three months, every one of the top identified people had received a phone call and a letter inviting them to either attend a seminar or to call an identified salesperson. A salesperson was assigned to follow up, and a manager was assigned to follow up on the salespeople. The total effort was accomplished through the part time efforts of two people. Is there any wonder why Cisco Systems has grown from nothing into a multi-billion dollar company overnight?

From my observation, the primary reason why Theory of Constraints companies have not made as much progress in marketing as elsewhere is bad assumptions and lack of willingness to market test different approaches. When dealing with the market and the reactions of large numbers of consumers, logic is only easy to establish after the fact.

You must talk to your customers and your competitors' customers — not through surveys, but directly. Use cause–effect logic to probe what it would take for them to buy more from you, not just to understand their complaints or needs. Try approaches that go against your gut.

Intuition in marketing can be counter-productive. For most of us, our intuition is based on a very small experience set, relative to the people we are marketing to. How can we ever hope to understand how other people think and behave? One suggestion is, the next time you're buying a magazine or a book, try picking one up from an area that you would never dream of having an interest in. If you hate hunting, pick up a magazine on hunting. If you detest psychology, try one of the books from that section. If you're a man, pick up a woman's magazine, and vice versa. Try reading some teen magazines. If you are like most people, you will learn that to be successful at marketing, there are some assumptions you must not make.

Summary

Before you can develop the right answers for marketing, you must surface and challenge every assumption you have about the market. You'll soon realize that you understand and relate to a very small segment of the world. Tastes, interests, values, and assumptions differ widely. Marketing is the challenge of finding which of the world's assumptions work best in finding clients for your products and services.

In the next and final chapter, I review the projects that must report to a CEO in order to secure the future of the organization.

19 Seven Projects that Must Report to the CEO

here are some things that a CEO must not delegate. I recently met a President of a computer chassis manufacturing company. He had attended one of my seminars and was excited to tell me about a recent change project in his company. "I personally devoted 40 hours a week to this project. Without my direct involvement, it never would have happened". He went on to explain how he had gotten everyone's commitment at the beginning of the project. Two weeks into the project, he had to "read the riot act" to get the effort ramped up to an acceptable level.

Every CEO I talk to has the same feedback. Without their extensive personal involvement, the change process would never have completed successfully. This chapter suggests critical projects that **must** report to the CEO. It is now decision time for you to secure the future of your organization. If you don't take action now, history says you never will.

Here are seven projects that I believe must exist in any organization to secure its future.

1. **Metrics and Rewards** — Someone with a good view of the forest and understanding of the organizational goals must review the entire organization's metrics (measurements and measurement systems) and reward systems from top to bottom. The right metrics will tell you not only where you've been, but whether you are headed in the right direction. Combined with the right reward system, metrics have a powerful influence on how people behave and on the extent to which you can improve your organization. A Theory of Constraints approach identifies the bad metrics and helps you find a solution with

no significant negative side effects. See the Acme Manufacturing case study for an example of the incredible impact on the bottom line.

2. **Enhancing Customer Value** — You need a full time team advocating for your customers — a team who can analyze the myriad of comments and complaints, interview customers in depth and come up with foolproof ways to increase sales by increasing value. This team will also experiment with different marketing approaches. In fact, they will be measured on the amount of experimentation they do, as well as the results they achieve through experimentation. See the Orman Grubb, Scarborough Utilities, and Health Services Company case studies for different approaches to enhance customer value with an impact on Throughput.

3. **Long-Term Competitive Advantage** — You need the very best people in your organization to find the one or two factors that will secure long-term loyalty with customers. This is not just a marketing issue. It crosses all functions and boundaries. Therefore, this team must be cross-functional and cross-process. For example, when the Japanese identified quality as the factor for the 1980s and 1990s, they made it a fact of life in all aspects of their business, from suppliers through engineering, production, delivery, marketing, dealer service, etc. The Japanese are now saying that quality is no longer the factor for competitive advantage. In the next millennium, quality is a necessity to be in the game. To gain competitive advantage, customization, and performance are the predicted factors.

The only way to identify the factor and be sure of the outcome is to have a lot of insight into your customers and their problems, and to have a detailed vision of what life would be like with your focus on "the factor". The Current Reality Tree is an excellent tool to understand customer problems and needs. The Future Reality Tree is a great way to visualize the new environment at a strategic level. See the Health Services case study for an example of an analysis of these strategic issues and a Future Reality Tree to address them.

4. **Human Capital** — Most people don't know how to think in a way that will give our organization long term competitive advantage and exponential increases in customer value.

I believe that the school systems are getting worse, which means that the schools and universities will not solve this problem. If you accept these assumptions, you must acknowledge the need to invest in the right kind of training — training that will help people think more deeply and more broadly about problems, solutions and the implications across functions.

Why must this responsibility report to the CEO? To answer, I can only offer my experience and my frustration. What I see, over and over again, is that without a CEO's directives and scrutiny, money is wasted and people do not develop. Without hard goals for development programs and training efforts, money is misdirected.

As an example, my partner and I teach a lot of people in public workshop formats, where you get one or two people from a company, mixed with several other companies. Throughout the session, we give out our fax number and assign homework, after the public session. We ask people to send their homework to us, yet no one has ever done so. In private sessions, dedicated to a single company with the CEO monitoring the progress, we receive homework frequently.

This does not mean that the public sessions are not worthwhile. It does mean that without hard goals attached to the training expense, and without the scrutiny of the CEO level, I believe that a lot of people attend workshops and never change their habits or contribute more to the organization. The training is not part of a total, strategic effort to develop people.

Another problem is that we rely too heavily on technology to solve our problems — the Internet, the computer, EDI, etc. The answers are not in the technology, but in how we employ it to provide better results for our customers. On this topic, most of the North American companies that I interact with are missing the boat. They are slaves, not masters, of the technologies they employ. See the ABC Forge case study for the power behind executive involvement in Training. See the Alcan case study for looking at human capital in another very important light.

5. **Policies** — In a seminar to a large high-technology company in February, 1996, Dr. Goldratt said that "I haven't seen an organization yet that has less than 70 wrong policies." Remember that bad policies, training, and metrics are the three diseases of every modern organization. Cure these, and the foundation for future security gets much stronger. Don't assume it's a one-time effort. A lot of today's bad policies were perfect 10 or 20 years ago. See the Scarborough Public Utilities Commission case study for examples of how changing procedures and policies on handling employees turned around their organization and is contributing to making them competitive.

6. **Chain-Wide Constraint Management** — Someone must examine the view of your market as a chain, from beginning supplier to delivery of the end consumer product, and identify the constraints blocking the chain itself from improving. If the constraints are within your company, this function's task to remove the constraint(s) is relatively easy. At least, it is easy compared to influencing other companies, particularly downstream from your company. However, these types of influences are what will build the intellectual capital of your company and increase its market value far beyond today's figures. Just read about the Skandia company and the number of relationships they have.* See the Acme Manufacturing case study for further insights into addressing issues of the supply chain.

7. **Chief Navigator** — Unlike the CEO, who is setting the direction, this person or team is ensuring that the ship is moving in that direction. It seeks the data necessary to confirm that the ship is making adequate progress in the right direction. It makes course corrections, and serves as a liasion between the board and the CEO, and other functions. It is also the chief communication function in terms of explaining the course that the captain has set, the progress being made and the challenges that lie ahead. This is not to detract from the communication from the CEO. Rather, it is to recognize that people in organizations need a lot of nurturing and information. Today, with the degree of change that organizations go through, this requirement is non-trivial. The CEO of a large organization should not be expected to handle this single-handedly. See the previous reference to Intellectual Capital.

Most executive teams are ridiculously short-staffed. Downsizing and restructuring have gone too far. Securing the future is not a passing fad. 7% of the 1995 Fortune 500 faded out this year. Your chances of fading out are increasing. The alternative takes a commitment of resources.

From the bottom of my heart, I wish you luck. My children and grandchildren are depending on your success.

---

* *Intellectual Capital: Realizing Your Company's True Value by Finding Its Hidden Brainpower,* Leif Edvinsson and Michael S. Malone, 1997, Harper Business.

CASE STUDIES

Introduction

The following seven case studies provide extensive examples of important
and strategic applications of the Theory of Constraints to securing the future.
The organizations profiled range from multi-billion dollar companies with
worldwide operations to small, single-owner manufacturing operations to
public utilities. The organizations are:

1. **Scarborough Public Utilities Commission.** An electric and water
    utility serving a population of 550,000 in a suburb of Toronto, Can-
    ada. The case study documents the CEO driven comprehensive anal-
    ysis of all chains within the organization and including suppliers, the
    identification of over a dozen core problems, and the subsequent
    transformation of the Utility to be competitive with the private sector.
2. **Orman Grubb Company.** A $35 million California furniture man-
    ufacturer. This case study illustrates the marketing and competitive
    impact of doing a TOC analysis on customer undesirable effects and
    conflicts, coming up with an irresistible offer to customers, and win-
    ning huge gains in terms of new customers, profits and competitive
    advantage.
3. **ABC Forge.** A large metalworking and heavy capital equipment
    steel forge company, with 300 employees, that has been in business
    since 1910. The case study documents how the executive team used
    the Theory of Constraints to address constraints internally and in the
    market, and how they changed employee measurements to revitalize
    a smokestack industry. While the company name and the names of
    the executives are disguised to protect their competitive edge, the
    analyses and results shown represent actual situations.

4. **A Health Services Company, in disguise.**   This case study reflects the TOC work I did several years ago. It shows how easy it is to identify major projects that will never yield benefit for the company. It also shows where and how to use TOC to build pillars of marketing strength that secure competitive advantage.

5. **Alcan,** a multi-billion dollar, worldwide producer of aluminum and its byproducts. This example of TOC shows how you can take two of the Five Thinking Processes, and use them in isolation to positively impact the lives of thousands of people.

6. **A High-Technology Company, also in disguise.**   The case study is a dramatic illustration of how predictable failure is, when you have executives who do not follow the "obvious" ideas required to succeed. It also shows that 80% of the work is not in finding a good solution, but in figuring out how to communicate it to gain buy-in.

7. **Acme Manufacturing Corporation,** a worldwide manufacturer and distributor of consumer products in the process industry. The case study documents the implementation of Throughput Accounting and Reporting and a complete change in measurements, leading to record sales and profits. The real company name and managers who implemented the Theory of Constraints have been disguised at the request of the Vice President, Finance, who is concerned about the competitive threat.

Case Study 1: Scarborough Public Utilities Commission

Scarborough Public Utilities Commission, or SPUC for short, serves a population of 550,000 people who live and work in this suburb of Toronto, Canada. In October, 1918, the first 80-watt street lights were turned on. The utility was born in 1920, as a non-profit organization overseen by a publicly elected board of commissioners. The utility offers both water and electric services to the entire population.

SPUC is the best example I have seen of a widespread implementation of the Theory of Constraints across all functions and levels in the company. In a two year period, the executive team has identified dozens of erroneous measurements and policies, and corrected them. Dozens of people have been trained in the techniques and actively use the techniques to chart new procedures, resolve conflicts, implement strategies and spread vision, mission and values across the organization.

The results in this not for profit environment are not as easy to measure as in a for profit organization. However, Kim Allen, the General Manager (equivalent to CEO) since March 1992 comments, "TOC is a significant factor to our business planning success. It has increased value to customers by more than $5 million annually. We have reduced Operating Expenses by $10 million and we are doing more work than we did a few years ago."

This value is easy to translate when you look at the history of rate increases before and after the Theory of Constraints approach to improvement. From 1990 to 1992, rate increases were almost 30%. Since 1993, there have been no increases. In fact, in 1997, a 7% decrease for industrial and commercial customers went into effect.

Measurements for reliability (outages and what that costs customers) went into place in 1993. In the high year, the cost to customers was $35 million. Last year, this was down to $10.7 million.

Kim initiated the first Theory of Constraints training process in August 1994. From our very first meeting, Kim expressed to me the urgent need for complete cultural change within the organization. Strategically, he explained, either they would learn how to adapt to private sector standards quickly, or they would eventually be replaced by competition.

The issues of change facing SPUC go well beyond quality. SPUC states, "The ability to provide products and services, even of very high quality, is no longer a competitive advantage. Today, nearly everyone can do this. The new reality requires agility to exploit the inherent flexibility and information content of new technologies to enrich customers in new and constantly improving ways. Instead of regretting or resisting constant change, we need to find ways to accelerate it as the resulting turbulence will stir up rich, new opportunities."

"Monopolies do not have the agility needed to compete in this environment. The new Utility must not be created by tweaking the old business paradigm, as appealing as it may be. It must leverage knowledge to operate profitably in a competitive environment of continually and unpredictably changing customer opportunities. The new Utility has perspective to see beyond the immediate and imagine a robust future providing speedy, relevant and flexible solutions to customers' unique problems."

"The determination of the businesses the new Utility should be in, the incorporation of the use of best practices and the creation of a new organization must be done using a thoughtful methodology. Alignment from task to vision statement is essential for high performance."

The first steps in achieving the alignment are answering:

- What to change?
- What to change to?
- How to implement the change?

The Utility is using a TOC master plan to bring their vision, mission and values alive throughout the organization. Here is what this document states:

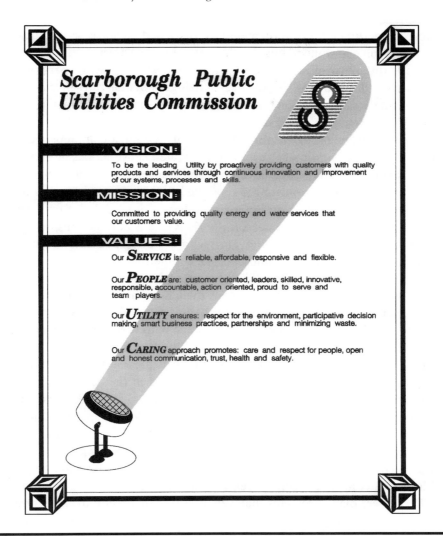

Figure 26   SPUC Mission/Vision/Values Statement

When I first met Kim Allen in 1994 I didn't fully appreciate the threat to Canada's public utilities. Until the late 1980s, they were pseudo government organizations, who were used to raising rates every year to cover cost increases. Many utilities, like SPUC, are unionized. In the late 1980s and early 1990s, as free trade began to hit the Canadian economy, Canadian manufacturers started looking to move their facilities to places like Ohio where the cost of labor and operations were much cheaper. Even within the greater Toronto area, a business spending a lot of money on electricity had a choice of 22 different utilities within a 50-mile radius simply by moving their operation a few miles away. The difference in rates between these utilities was growing, particularly since some utilities, like SPUC, were subsidizing residential rates by payments from businesses. In addition, cheap natural gas became an attractive alternative to electricity for many consumers.

Of course, when businesses move out of your area and are not replaced, you must either reduce your cost structure or raise rates. However, raising rates was becoming more and more of a threat to keeping business in the area. Since the Mayor of Scarborough is one of the appointed commissioners who sits on the governing board of SPUC, you can imagine that there was a lot of pressure to contain costs and not lose the tax contributions of Scarborough businesses.

Kim had some other insights. He saw a future where there would be competition, potentially even international competition, to provide electrical services to a community. After all, the power is distributed from a wire going into a building or home. Who says that the source has to be local? In fact, in the near future, there is speculation that even the physical wire may not be necessary. Look at the progress being made with fuel cell and small scale gas generation distribution systems.

Kim reasoned that one of SPUC's strengths is the fact that they have the infrastructure to service every home and business in the community. With the functions and information systems necessary to do such things as bill for services, provide customer service (telephone inquiries, repair, maintenance, start and stop services, etc.), Kim felt that it made sense to use this infrastructure for more than just electricity and water. In fact, why not use the infrastructure for any kind of utility service (e.g., natural gas, district energy, cable and satellite TV, telephone services, internet, etc.).

There were many problems that were blocking this vision from becoming reality. For one thing, the utility was not competitive with the private sector in terms of their cost of doing business and providing service. For another, the employee culture lacked sufficient empathy for the customer to be a world-class competitor. The labor relations were not as good as they needed to be, given trends towards downsizing in the neighboring utilities. The union was still operating with a "business as usual" attitude when, in fact, the economic climate was about to change drastically.

In fact, in the Current Reality Tree analysis of SPUC, 19 root problems were identified (see "Root Problems," below). All of these problems needed to be addressed, or at least underway, before SPUC could even dream of adding new services.

Many of the changes that SPUC needed to make in how they managed and used the labor force were being blocked by the union. Reporting to a publicly elected political body doesn't make things easy either.

What first intrigued Kim about the Theory of Constraints was the link between strategies, tactics, and implementation plans. As he said, "Our management team, for the last several years, has gone to annual planning meetings and come back with lists of 75 new ideas to check out, and 150 different things we should do. What we found the following year was, very few of these ideas had been implemented and the new list became even longer."

Kim saw the Theory of Constraints as a complete, integrated methodology that would consider the relationships between functions and processes. He viewed it as a means to link strategies and tactics across the entire organization. He wanted to focus the team on eliminating the key or root problems that were a prerequisite to getting SPUC positioned to be competitive with the private sector.

If you are considering major changes in your culture, your strategic and/or tactical approaches, listen carefully to what Kim says about such an effort:

"For other executives starting down the Theory of Constraints path, I offer the following:

- TOC is not an academic exercise. To be successful, it requires involvement, support, and the use of the process [to your real life problems].
- TOC cannot be totally delegated. The scrutinizing of the work by the executive is a critical part of the success of the tool.
- The CEO and his or her direct reports must be fully trained in all of the Thinking Processes of the Theory of Constraints. TOC is driven top-down, with the participation into the solutions coming from the bottom up.
- Teach the lingo and the day-to-day applications of the Theory of Constraints to the Level 2 reports. Otherwise, they will get lost in the lingo and not support it if they do not understand it.
- Develop in-house experts in the Theory of Constraints, not on the executive team, who will stay on top of the process and check you out from time to time. For example, at SPUC, the efforts of Wendy Howse, an expert in all of the TOC thinking processes, is essential to our success. In addition, Daria Babaie was trained in several of the TOC processes. His relentless work in developing our business planning processes made many of our results feasible.

- Use the Transition Tree for some day-to-day activities so that people throughout the organization become familiar with the lingo and see the value of the process (for procedures, see examples that follow.)
- Don't get caught-up in the discipline of the process! Implement once you have a "good" solution (i.e., not perfect). If we did it perfectly the first time, where's the room for on-going improvement?
- Don't forget the importance of buy-in and communicating it to all who are involved in the process.
- Analyze to make sure you are working on the root problem. Do not assume that you have the solution and it just needs to be implemented.

What Kim discovered was that TOC is the only methodology that provides a clear picture of how long and how involved implementing real change takes. "There are no shortcuts, even though we continue to try. Weakness occurs in the areas of buy-in and communications. This continues to cause activities to be less successful than they should be."

SPUC has worked on breaking five major constraints to their future security and success:

1. **The constraint that people will do what they want to do unless there are consequences.**   This constraint implied a combination of training, measurement and procedural problems across the entire organization that needed to be corrected. Breaking this constraint made is possible to begin to align activity [between and within functions], delegate work and assign accountability.

2. **The collective agreement was viewed as a constraint to disciplining people — providing the consequences if they did not behave appropriately.**   [My observation is that the executive also learned a great deal about when to discipline, when to train and when to take other action in response to inappropriate behavior, i.e., discipline is not always the answer, but in any case, you must understand the cause of the inappropriate behavior before taking action. This simple realization was never documented in a procedure before, and was not obvious to the supervisors and managers trying to enforce discipline].

   The development of the "Inappropriate Behavior Transition Tree" has changed how SPUC approaches discipline. It involves the employees and the union (UWC — Utility Workers of Canada) at the early stages. Since its inception, every termination has stood up to the arbitrator's decision.

3. **Customer rates did not match the wholesale cost of their products.** This was the cause of a raft of problems, resulting ultimately in the loss of commercial customers and a loss of competitiveness. The new rate structure more closely aligns costs and provides for open competition. This was a major accomplishment in a political environment where adjusting rates meant that there could be short term winners and losers, all of whom are voters.

4. **The business planning process did not lead to a process of ongoing improvement.** This was a problem of skills, policies and procedures and measurements that was corrected using a fully automated and integrated set of processes and software. A combination of Throughput Accounting (using T, I, and OE) and Activity-Based Costing is part of the planning process.

5. **Staffing levels were too high.** [My observation is that the TOC process got started too late to keep everyone employed in the short term. If the utility had been positioned earlier as an organization on a mission to handle a wider range of services, it might have been possible to avoid attrition. Nonetheless, TOC helped enormously.]. Using TOC, people were redeployed to avoid layoffs or expensive buyout packages. In 1992, SPUC's staffing level was 550 people. Today, it is 421 — all accomplished through natural attrition (people retiring, people leaving to seek other challenges as SPUC encourages people to become their personal best, and a few through termination resulting from the Inappropriate Behavior Transition Tree (see Figure 27).

Kim Allen's initial team consisted of Kim and his direct reports:

- Joe Bailey, Director, Electric Division
- Brian Doherty, Director, Water Division
- John Dunnett, Director, Customer Services
- Jim Tearne, Director, Corporate Services

The team met once per month for 3 to 4 days, to learn the Theory of Constraints processes and apply them to SPUC's situation. In between sessions, the team would meet and review their work together, and help ensure that the various analyses made sense to the entire team, according to the rules of logic.

As the team began to implement their first injection (idea) from their Future Reality Tree, they hit a brick wall. The level of resistance to change and to the analysis that the team had completed was enormous. The fact that each member of the team was using a new language (the language of TOC — trees, clouds, etc.) created a mountain of suspicion among the employees.

The team decided that one of their intermediate objectives, which was skills training for all of their managers and supervisors, was out of sequence. They had envisioned it as a requirement, but occurring much later in the change process. In fact, it became a prerequisite for the implementation of any significant change across the organization. As a result, Kim decided to pilot a program training 16 people in the Theory of Constraints techniques as applied to day-to-day issues. The 16 people were selected from various levels and groups in the company. They included, for example, the Union President, managers, supervisors, engineers, etc. My partner, Jackie and I, performed the initial training. We were first hand witnesses to the combination of excitement and skepticism that existed within the organization. We also realized the opportunity, with the more open attitude of the new union president, for bringing about change that would have enormous benefit to all union members.

With the success of the pilot, Kim decided to train an internal consultant, Wendy Howse, to learn the entire Theory of Constraints Thinking Processes. Wendy became the in-house expert, available to groups to apply TOC to significant problems, to aid implementation of change and to perform all further in-house training. Since Wendy's initial training, she has facilitated workshops and developed several dozen people within SPUC and consulted on many difficult problems. The dedication of her time and her expertise undoubtedly help account for the breadth of usage and success with the methodology within SPUC.

In hindsight, Kim says that he would have done it differently. "I would have selected fewer options to implement and got them in place before tackling others. Only about 5% of the people are prepared to drive change. I overloaded them rather than getting a large percentage of them on side with the change effort".

These comments are remarkably frank and insightful. Remember that the Theory of Constraints is already a focusing tool. Picture that even after focusing from hundreds of problems down to 19 problems, people were still overloaded with the planned change exercises. Compare this to organizations that are not focused on core problems. Do you begin to understand how hopeless some of these organizations are at ever achieving a strategic victory? Perhaps their only saving grace is that their competitors are equally encumbered.

Throughput that Recognizes Customer Value

In a not-for-profit organization, how do you measure Throughput? SPUC felt that the concept of Throughput was very important, as a focal point for everyone's efforts within the utility. At SPUC, it is defined as the value added for the customer.

$$T = \$in - \$wholesale\ cost + \$additional\ customer\ benefits$$

where:

$\$_{in}$ is the money SPUC receives from billing revenue, selling services, return on investments, etc. less the cost of collections, credit extended and writeoffs.

$\$_{wholesale\ cost}$ is the cost to buy water and electricity

$\$_{additional\ customer\ benefits}$ are the benefits that a customer receives beyond the ones already paid for in $\$_{in}$. For example, improvements in customer interruption costs, energy management savings, reduced waiting time, etc.

If SPUC were strictly measuring customer value, rather than use *all* of the dollars coming in to them, they would use the difference in rates between their utility and neighboring utilities. This is part of the real value of buying electricity in Scarborough vs. metropolitan Toronto, for example.

This concept of real customer value is important to every organization. For example, if power is interrupted and a business must shut down its manufacturing or its retail operation, this may impact customer throughput. Also, the extent to which a customer must spend money to protect themselves from power interruptions increases their operating expenses. Therefore, a utility's efforts to improve in these areas adds real, tangible value to the customer.

Every business needs to measure how much real value they have added to their customers in the past year.

The Inappropriate Behavior Procedure (Transition Tree)

Any new methodology creates tension and fear. Employees openly wondered why the executives were locked up in this classroom with these mysterious "Theory of Constraints" consultants. Were they plotting some restructuring or downsizing? The mood in most organizations that I visit is one of skepticism.

Therefore, any new methodology must overcome a fantastic amount of resistance before acceptance.

The Inappropriate Behavior Procedure, from my point of view, became a turning point for overcoming resistance. It is intended as a procedure to deal with any incident or accident with employees taking action that was deemed inappropriate. Examples include stealing, coming to work late, using company equipment for personal use, authorizing unnecessary overtime, etc.

Since this became a way of getting union, employees, supervisors, and managers on the same wavelength, and of getting everyone in the organization familiar with the lingo and the approach of TOC, I've included it here practically in full. It is a model that can be adapted by any organization, and I am indebted to Kim Allen for his willingness to share it.

I recall that when the executive team began constructing this tree, they couldn't fully comprehend why there was such disparity across the organization in the way these incidents of employee behavior were handled. They already had a procedure and training. Yet, some supervisors were not following through at all, while others were creating situations with the union that were inflammatory.

The executives' first attempt was about one and a half pages long. It took leaps of blind faith, using that document, for a supervisor to understand what was expected of him. The final outcome, while not an absolutely perfect tree by academic standards, has been proven "good enough" to withstand the test of usage across the organization over a long period of time.

The term "DRP" is an acronym for "Directly Responsible Person", meaning the boss or supervisor of the employee who is behaving inappropriately.

The Transition Tree begins by instructing the supervisor on how to collect initial information and investigate the incident. The collection of facts is very important. The way to determine the cause of the employee's inappropriate behavior is detailed, as well as the appropriate and very different actions to take, depending on the cause.

The first five pages of Transition Tree (Figures 27 to 33) detail the generic procedure. The last two pages show how to deal, specifically, with an employee who is blocked from behaving appropriately by a policy or by the collective agreement or by his or her own lack of skill (incompetence). In either case, the objective is to eliminate the root of the problem. Of course, if the employee is competent and is not blocked by any policy or the collective agreement, and still behaves inappropriately, then disciplinary action is in order. Alternatively, in this case, if the employee disagrees with the disciplinary action, they may file a grievance.

Note that the level of detail shown in the procedure was not overkill, even for very experienced supervisors, since there was so much inconsistency in how the procedure was being carried out. However, with the new procedure, which was opened up for scrutiny by the union and employees, even a new supervisor could handle such incidents very effectively. The full procedure follows:

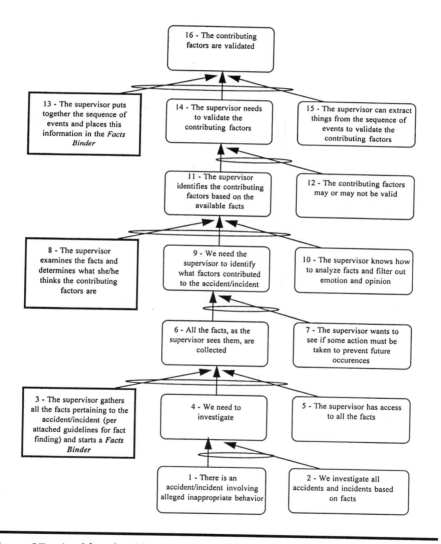

Figure 27   Accident/Incident Inappropriate Behavior Transition Tree #1

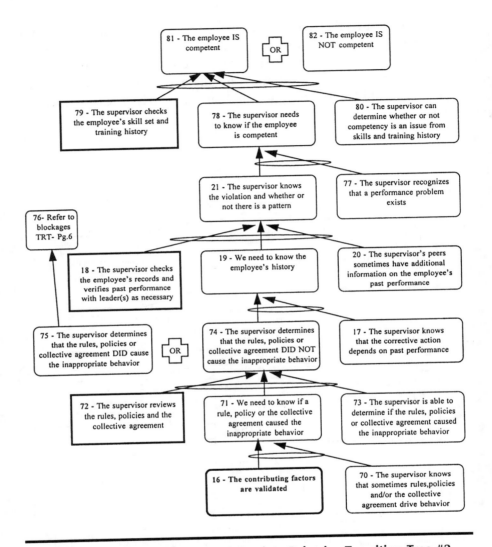

Figure 28   Accident/Incident Inappropriate Behavior Transition Tree #2

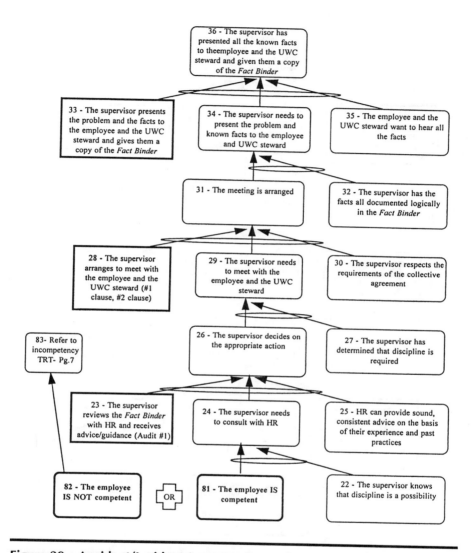

Figure 29   Accident/Incident Inappropriate Behavior Transition Tree #3

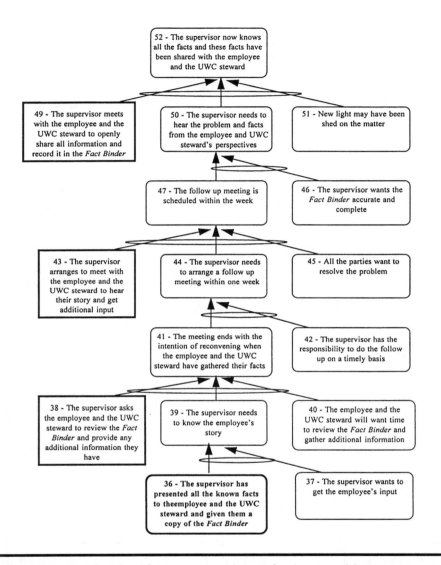

Figure 30   Accident/Incident Inappropriate Behavior Transition Tree #4

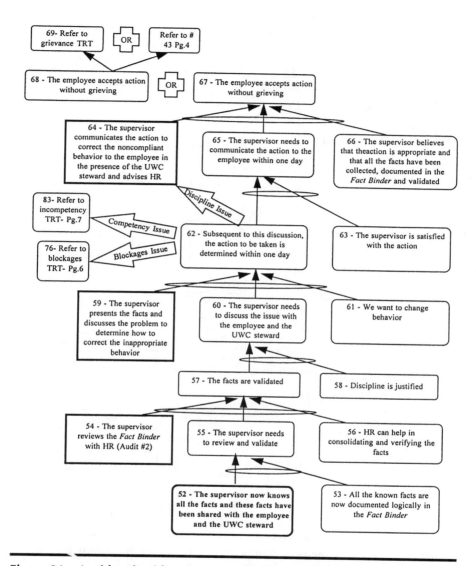

Figure 31 Accident/Incident Inappropriate Behavior Transition Tree #5

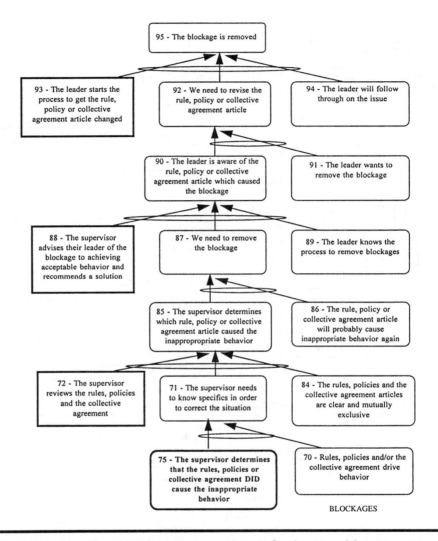

Figure 32 Accident/Incident Inappropriate Behavior Transition Tree #6

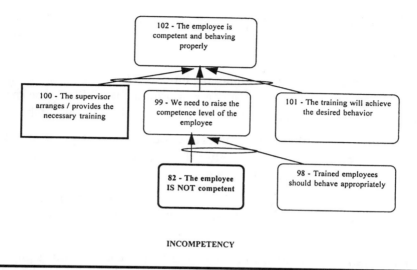

INCOMPETENCY

Figure 33   Accident/Incident Inappropriate Behavior Transition Tree #7

SPUC Theory of Constraints Analysis

SPUC identified 35 chains within their organization. A chain is a set of interdependent events (procedures, processes, actions) necessary to produce a result or meet an objective. Within the 35 chains, 19 root problems were singled out.

To help with the cross-referencing between chains, SPUC invented a simple, alphabetic code to signify each chain. Here are the identified chains and a general description of the area covered:

A.   Rates — deals with the impact of rate structures for billing water and electricity services, and the impact those structures have on competitive positioning and the ability to retain customers.

B.   Account Management — specific to how customer accounts are handled relative to extending credit and the collection processes.

C.   Corporate Support — relates to all of the corporate support services (e.g., Information technology, management reporting, performance tracking).

D.   Budgets — describes the problems blocking an effective business planning process.

E.   Owed a Living — describes the interrelated events and attitudes leading to undesirable employee behavior, abuses of privileges and deterioration in customer value.

F.   Status Quo — this tree examines the human behavior and management attributes that block a process of ongoing improvement at SPUC. The chains of communication and management were both involved in this analysis.

G.   Job Postings — this tree examines all entities that block getting the best people for a job and also result in after-the-fact fire fighting and employee disillusionment with executive decisions.

H.   Commission — the commission is like a board of directors in a company. This tree examines the relationship between the executive and the commission, and the issues that block progress.

I.   Environment — deals with the chain to keep the community's physical environment safe and clean, the costs involved and the interactions with other government agencies.

J.   Behavior — deals specifically with measurements, performance expectations and their communication across the organization. This tree explains why good people behave badly.

JA.   Leaders — describes the chain of leadership coaching and further explains why some people are not capable of doing the job they are in.

K.   Customer Interruptions — describes the interrelated factors that block meeting customer outage improvement targets.

L.   Completion of Work — otherwise known as project or program management, this tree explains why many programs are not successfully completed.

M.   Wholesale Costs — describes the supplier chain and government influence on costs

N.   UWC — deals with the union relationship at an organization to organization level

NA.   Union Stewards — deals with how union stewards interact on the job on issues in a way that blocks SPUC from achieving their goals.

O.   Stress — every organization has a chain that deals with managing and implementing change and the extent of teamwork involved. In the case of

SPUC, the examination of this chain led to this current reality tree describing the extent of the stress problem and its root cause.

P. Our future — because SPUC is a not-for-profit, pseudo government organization, it's future is not entirely in the hands of the management team. In fact, the Ontario provincial government has legislated the amalgamation of all Toronto utilities and municipalities into one entity. This tree examines what blocks SPUC from controlling its future.

Q. Communications system — describes what is blocking SPUC from taking on a major new utility support function of the information superhighway.

S. Independent Action by executives — within the executive chain, this tree examines what blocks executive alignment in working towards the overall goals of SPUC.

T. Teams — given that most change is implemented by teams, this tree reviews the root problems blocking teams from meeting their objectives.

TA. Mutual Gains — this tree examines how a lack of skills and practice in dealing with problem solving leads to a lack of effort to find solutions that mutually benefit both parties in a conflict or problem situation.

TB. Labor Relations — deals with the chain of problem management and escalation.

TE. Inter-departmental barriers — deals with the practice of local optima — each person/manager works for the success of their department or function, which often conflicts with working with a utility-wide perspective.

TF. Benchmarking — determines the root problem behind unfavorable differences in cost structures between the utility and the private sector.

TG. Staff Perception of Executive Team — the whole communication culture between executives sets an atmosphere that the rest of the employees are acutely aware of.

TH. IT Performance — this is the Information Technology — services and support — tree.

TJ. Utility Industry — examines the relationships between utilities that lead to wasted resources.

U. Long range plans — examines the factors blocking long range increase in Throughput.

V.   Surplus property — SPUC has a lot of surplus property that increases OE.

W and WA.   Ontario Hydro — Ontario Hydro is the supplier of wholesale electricity to SPUC. This tree examines the aspects of the relationship with this sole supplier that blocks SPUC from achieving their goals.

X.   Metro — Metro is the supplier of wholesale water to SPUC. This tree examines the aspects of the relationship with this sole supplier that blocks SPUC from achieving their goals.

Y.   Customer Needs — Customer service and support chain is examined in this tree.

Z.   Costs — All elements of costs, cost reporting and being competitive with the private sector are assessed in this tree.

From all of these current reality trees, the following root problems were identified (see Figure 34).

These root problems account for hundreds of undesirable effects across the organization. The letters correspond to the chains described above. The numbers signify a specific entity on that tree from which flows all of the undesirable effects. Many of the root problems described above cause undesirable effects in other chains, not just in their originating chain.

Let's look more closely at one of the Current Reality Trees — Tree D (Figure 35) dealing with the topic of budgets.

In this tree, we're showing that several root problems are responsible for the problems the organization has related to budgets. D16 — "Our decision making process does not require people to use a common set of measurements (MDM)," is one root. However, notice A26 on the left side. This root problem was identified in the "A" Current Reality Tree on Rates. The entity above it, C22, was identified on the Corporate Support Current Reality Tree. And entity J111 on the right was part of the tree dealing with Behavior.

We can also see that entity D16 is a root problem for many other chains in the organization, since it flows into trees S, V, Z, and T. It's effects also flow out to tree U and Tree TF.

The assumption on a Current Reality Tree such as this one is that problems stemming from other trees will be dealt with by addressing the root problems on those trees. When the root problems of those chains go away, then entities such as F3 in the upper right — "Managers are very protective of their existing turf," will also go away.

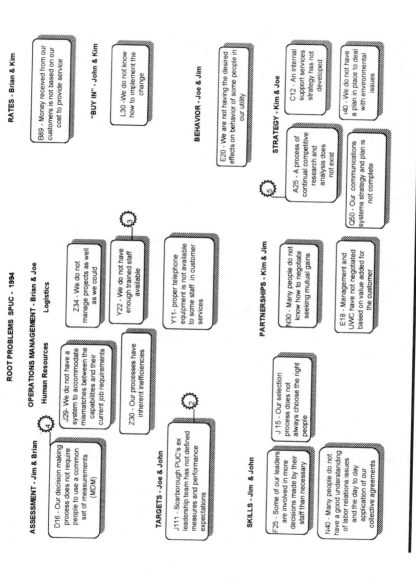

Figure 34 Current Reality Tree of SPUC Root Problems (1994). UWC refers to the Union. PUC stands for Public Utilities Commission.

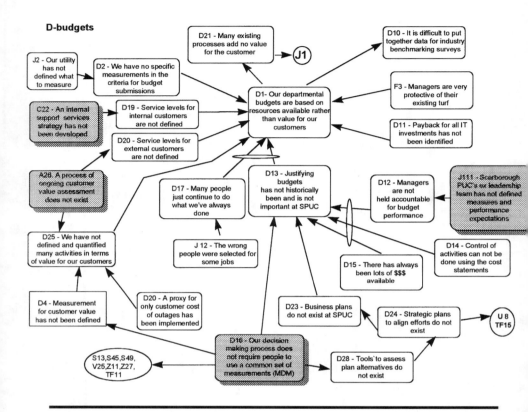

Figure 35   Tree "D" Dealing With Budgets

The root problem D16 was not something new to SPUC executives. The surprise was how it was tied in to so many negative effects. The issue of measurements was one that troubled SPUC for years. Here is the evaporating cloud that described the dilemma and the assumptions behind it. Also, from this conflict diagram (Figure 36) you see the beginning of a solution — possible injections to overcome the core problem.

When you relate this conflict diagram to the Current Reality Tree, you can see the two sides of the conflict running up the tree, almost as two branches of the tree. On the left branch of the tree, you can see the emphasis on providing customer value and the elements that block it. On the right side, the elements blocking the ability to meet the interests of the utility and performance to budget are represented.

The executive team tried to uncover assumptions and possible injections under each arrow of the conflict. For example:

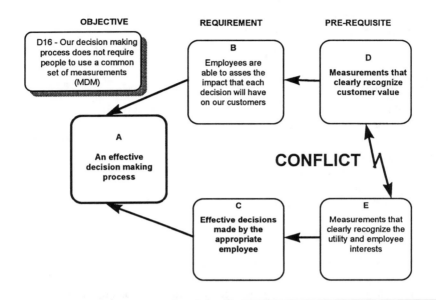

Figure 36 Conflict Diagram of Root Problem D16. Note: The author has taken some editorial leeway in simplifying the original contents of this conflict diagram. Assumptions follow.

Assumptions: A to B

- Resources are limited.
- The highest priority work must be done first.
- Decisions made must be aligned with our goal.
- Decisions are often challenged by customers and the commission.
- Consistency of application is important.

Assumptions: B to D

- Customers want maximum return for their money.
- Measurements must be understood by every employee.
- Employees care about customers.

Assumptions: A to C

- Execs have limited time to review proposals.
- The person closest to the action has the best info to make the decision.
- Many decisions must be made ont he fly.
- Decisions must be made using fact rather than personal preference.

- Bureaucracy gets in the way of good customer service.
- People want to make decisions.
- Our current decision making process is too slow.

Assumptions: C to E

- Employees currently assume that empowerment means do what you want.
- Employees assume that measurements are not clear and managers are the one's who should make decisions.
- Employees must know what decisions they are expected to make.
- Some people will take ineffective action without a specific direction.
- Employees will invent their own tools if none are provided.

Assumptions: D to E

- Employees don't think they will be supported if they make a decision (i.e., they perceive that it is impossible to meet both requirements).
- Doing your own think may not be beneficial for customers.
- We don't know how to balance the needs of individual customers versus the impact on the rest of the customers.

They broke the conflict under the D to E conflict arrow by overcoming the assumption that "We don't know how to balance the needs of the individual customer vs. the needs of all the customers". The opposite of the assumption, which is that "we do know how to balance all of these needs" was accomplished by going to a common measurement system using T, I and OE parameters. Figure 37 is what the Future Reality Tree looked like with this injection.

From this Future Reality Tree, it looks like everything is accomplished from just one injection at the bottom of the tree — Box A1 — stating that "We use T, I and OE as our measurement for decision making". From an academic point of view, we could examine the tree and find many insufficiencies — effects that require other injections or entities to cause those effects to occur.

In this case, you must remember that you are seeing just one of 35 trees. There are many interrelated cause-and-effect relationships, and therefore many injections coming from different sources. For example, Figure 38 is another part of the budgeting Future Reality Tree indicating injections from other sources.

In this case, it is assumed that Kim and Jim have implemented injections causing certain necessary conditions to be in place, such as E20 — "We are

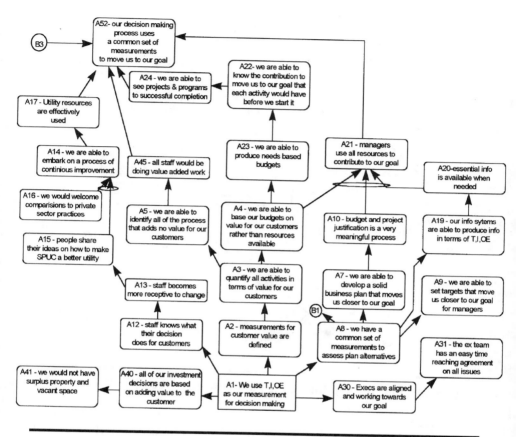

Figure 37   Injection of T, I, and OE Parameters

dealing with inappropriate behavior". This is an effect that came out of the Future Reality Tree dealing with the E chain described earlier ("Owed a Living").

Once work had progressed on Future Reality Trees, the team needed to take each injection, assign responsibility for managing the project to achieve the injection, and perform detailed planning. For example, you may recall that one of the chains deal with Corporate Support Services. Figure 39 is what the Current Reality Tree looked like.

The root problem was the lack of an internal support services strategy. On the Future Reality Tree, one of the injections was "We have customer-focused support services." A list of obstacles and intermediate objectives became a Prerequisite Tree, which is shown in Figures 40 and 41.

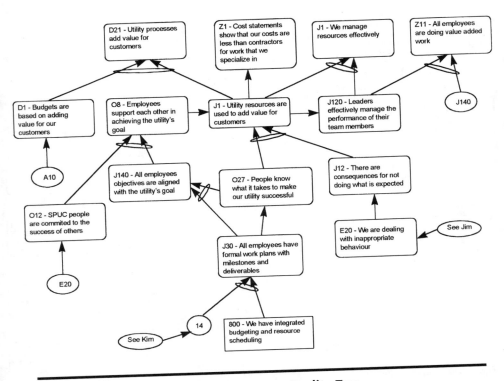

Figure 38   Injections in Budgeting Future Reality Tree

What is evident from these examples is the level of commitment of managers organizationwide that was required to embrace the techniques and make them a standard part of the planning and implementation processes at SPUC.

This result happened with a lot of bumps along the road and a great deal of resistance. TOC was helpful was in tying together the strategies (what injections or ideas were the right ones — what desirable effects were the team looking to accomplish to reach their vision) and the tactics (what actions do we need to take to achieve the injections).

SPUC team leaders used the Transition Trees to assign individuals to actions and intermediate objectives, and to track progress along the way. Rather than the usual form of project management, which often tracks only the actions taken and not the results, the Transition Tree is used to track results and to understand, through the defined logic, why results were not achieved from planned actions.

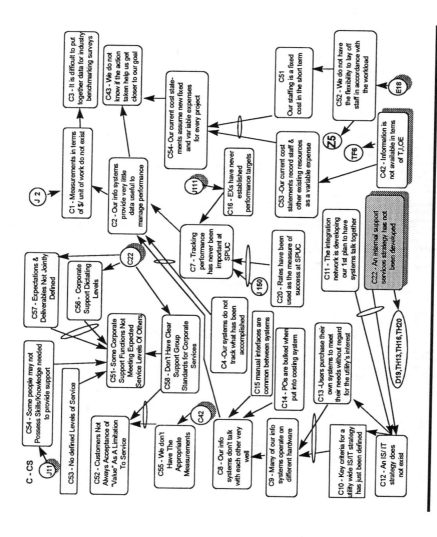

Figure 39   Chain Dealing with Corporate Support Services

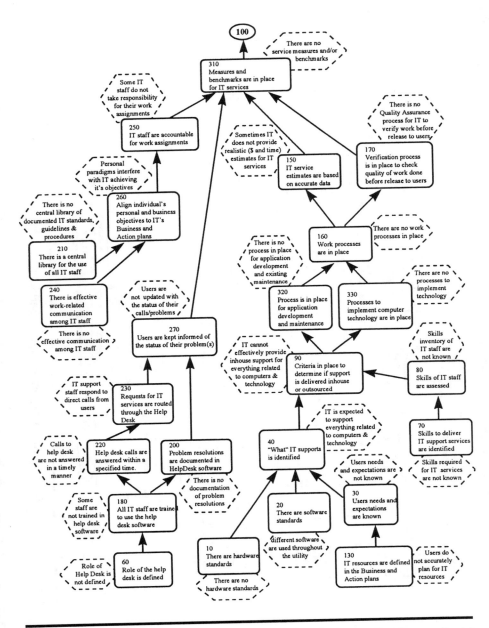

Figure 40   Prerequisite Tree for Customer-Focused Support Services — Part I

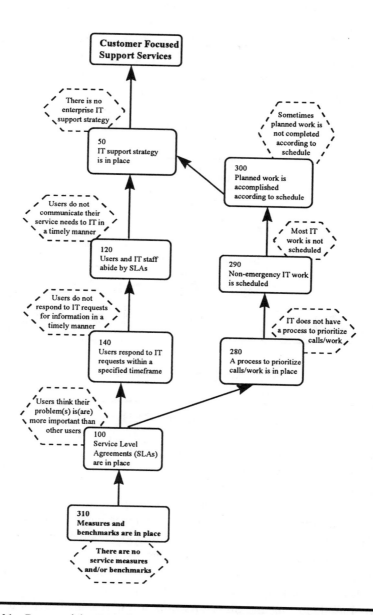

Figure 41   Prerequisite Tree for Customer-Focused Support Services — Part II

Since the initial Theory of Constraints experience, the tools are being used to solve a wide variety of day-to-day problems, and to develop procedures for difficult or sensitive tasks. For example, what action do you take in a unionized environment when an employee is absent for more than five consecutive days? Here is the answer, including instructions to supervisors on how to read the Transition Tree (see Figures 42 to 46):

How to Read a Transition Tree (TRT)

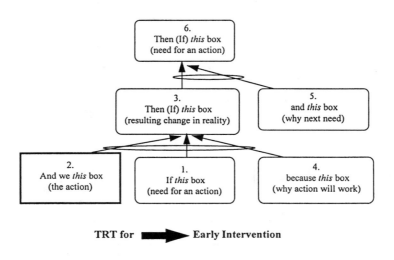

Figure 42   How to Read a Transition Tree (TRT)

Anyone who had already taken training in the Theory of Constraints tools would not require these instructions. However, the training was never forced on employees. Therefore, the assumption is that there will always be people who can use these procedures without having training in how to construct them.

From procedures like this one, every employee has the ability to see not only what actions management will take in a given situation, but why they will take those actions and what results they expect.

The TOC methodology is used in conjunction with other techniques such as TQM, Benchmarking, Balanced Scorecard, Analytical Hierarchy Process, Activity Based Costing and others. As Kim notes, "A skilled carpenter can do wonderful work with a hammer and saw. A handyman can do even better work if he uses the custom tool for the job. TOC provides the best overall framework for developing the blueprints and plans. Other tools are better at some [specific] areas."

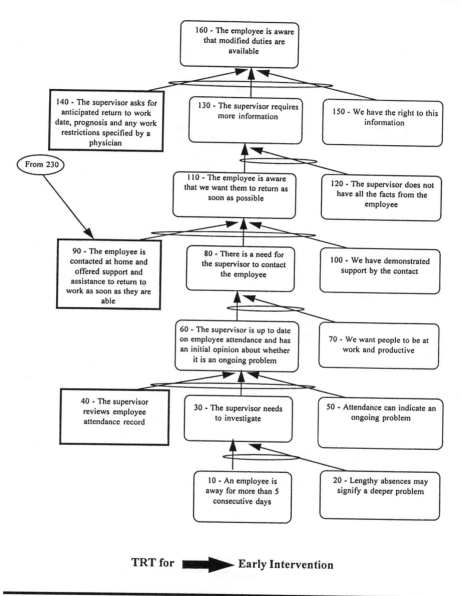

TRT for ➡️ Early Intervention

Figure 43   Page 1 of SPUC Transition Tree

Kim is very adamant that the CEO must be proficient in the Theory of Constraints to get the benefits. "Using TOC is a lot of work, and requires discipline to use. It would have fallen by the wayside if I were not a Jonah [expert in the Thinking Processes of TOC] and pushing the Directors to use it."

I couldn't agree more.

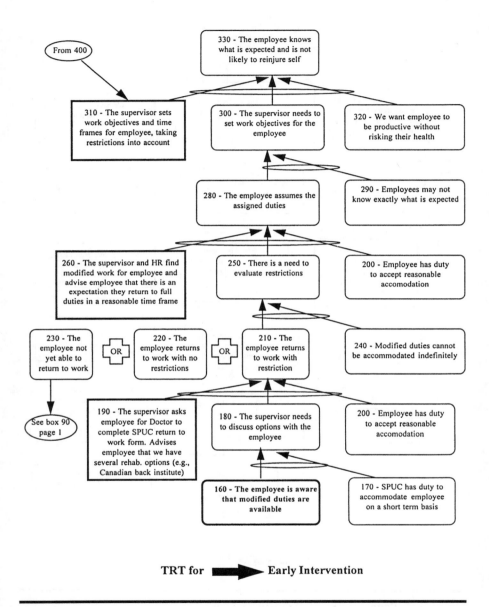

TRT for ➡ Early Intervention

Figure 44　Page 2 of SPUC Transition Tree

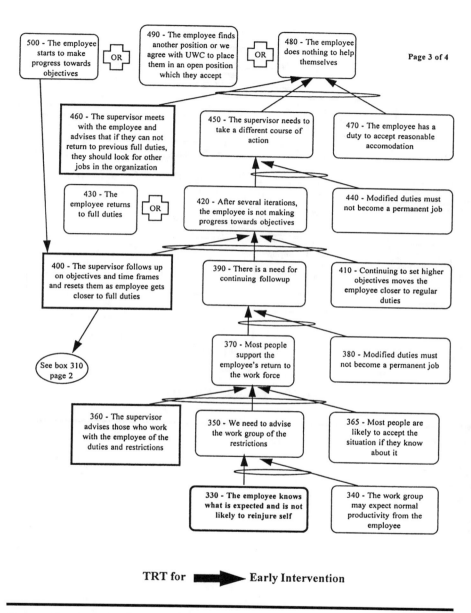

Figure 45   Page 3 of SPUC Transition Tree

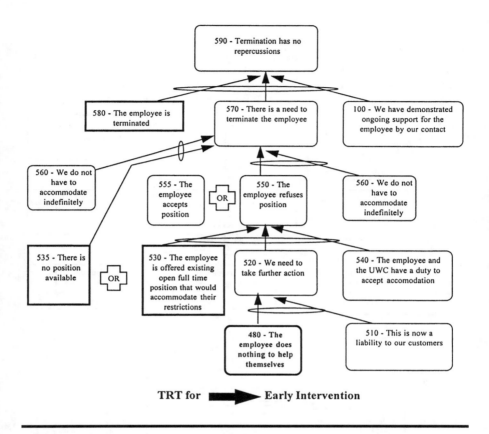

Figure 46   Page 4 of SPUC Transition Tree

Case Study 2:
The Orman Grubb Company

Orman Grubb is a $35 M California manufacturer of contemporary oak, home office, home entertainment and bedroom furniture. Jeff Grubb, the CEO, started the business in a garage 20 years ago.

Jeff jokingly (?) claims that his industry is noted for having the greatest number of undesirable effects. Certainly, it has the simultaneous challenges of dealing with all of the problems of manufacturing, transportation and distribution logistics, and retailers who go in and out of business constantly.

Jeff's first exposure to the Theory of Constraints came from reading *The Goal*, Dr. Goldratt's bestseller about Manufacturing Constraints, in 1990. He began to apply the concepts directly to his plant. In 1993, he attended a TOC seminar on Sales and Marketing and was trained in the Thinking Processes of TOC in 1994.

Orman Grubb was particularly interested in breaking their market constraint. They knew that they had the plant capacity to handle many more customers. The question was how to cost effectively market, win, and outsmart their competitors for a long time into the future.

The concept of coming up with an irresistible offer to current and prospective customers by overcoming the undesirable effects in the customer's environment appealed to Orman Grubb. They already had a lot of intuition and suspicion about what those problems were. They began further research by talking to a combination of customers, salespeople and other internal people.

As Jeff noted, "Be careful of the customers you collect UDE's [undesirable effects] from, because some of them have special agendas." Many customers provide a long list of every problem and fantasy they have, without distinguishing which ones, if fixed, will cause them to buy more from you. Jeff suggested that it is essential to use internal people, sales people and people you trust, who don't have their own agenda.

Core Problems of Furniture Dealers

As with any entity, furniture dealers have several interdependent chains within their organizations that drive their business. For example, every furniture store deals with sales and marketing, purchasing, finance and cash flow, administration, inventory control, etc. Each chain has one or more core problems.

From Orman Grubb's point of view, you should not even try to solve all of their problems, for two reasons: (1) you don't have the ability to solve *all* of them and (2) you need to hold back some solutions so that when competitors catch up, you have your next moves ready.

In his TOC analysis, which involved many conflict clouds, Orman Grubb resolved three clouds that represent the core problems of dealers and why they are core problems. As is typical of core problems, these are issues that dealers have been grappling with for years, without resolution.

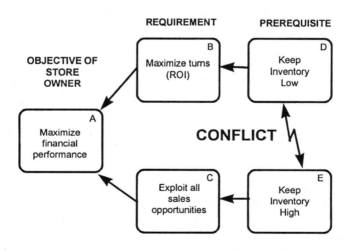

Figure 47   Cloud 1 — Matching Store Inventory to Consumer Demand

Store owners want to maximize financial performance. In order to do that, they must strive for the maximum inventory turns. In order to maximize inventory turns, they must keep inventory low. On the other hand, in order to maximize financial performance, they must exploit all sales opportunities. In order to do that, they must keep inventory high.

Jeff explains that dealers typically bounce back and forth between low and high inventory, never resolving this conflict but compromising — keeping higher inventory in the seasonally high buying periods and lower inventory at other times. Graphically, the picture looks like the following:

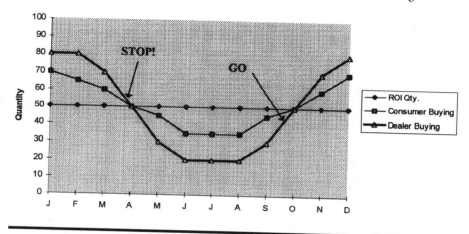

Figure 48   Seasonal Inventory Highs and Lows

The months of the year are at the bottom of the chart. The middle line indicates the minimum sales quantity required to meet the store's ROI requirement. Consumers buy seasonally. The consumer buying pattern shows high sales at the beginning of the year that drop off in the spring and summer and start up again in the fall.

Store buyers tend to overreact to consumer buying habits, by buying huge quantities in anticipation of seasonal buying, and then panicking and cutting all purchases until their inventory drops to a critical stage.

These buying habits create enormous pressure on the manufacturer, who has to work out the transportation and manufacturing logistics to handle the peaks and valleys. One way that is typically used by manufacturers is to quote long and variable lead times to dealers. This helps them buffer the "noise" (i.e., the fluctuations in demand) in the system.

However, consumer buying habits hurt store owners. Store owners become victims of the Pareto Principal — the 80/20 rule:

- 20% of the merchandise in a store is best selling.

When the dealer stops buying, 80% of the inventory reduction is in the 20% of the merchandise that's best selling. Because many consumers wish to purchase for immediate delivery, the dealer is out of stock on the popular merchandise and loses sales when they can least afford it — during the period of slower consumer buying.

Ideally, retailers would love to buy small quantities of furniture frequently to match consumer demand. The following cloud explains the conflict that blocks meeting this need:

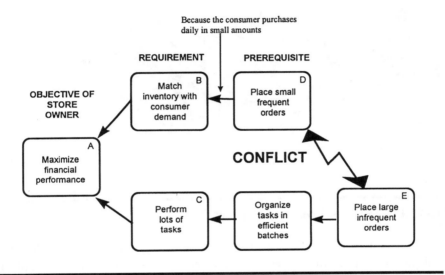

Figure 49   Cloud 2 — Buyer's Cloud of Order-Analysis Time

In order to achieve financial performance, the store buyer must match inventory with consumer demand. Ideally, to do this, they must order frequently, in small quantities, because that is the way that the consumer purchases. On the other hand, to maximize financial performance, the store buyer must perform lots of tasks. They don't deal with only one supplier, but with many and there is a lot of data to consider. To get their work done, the must organize their tasks into efficient batches. When it comes to buying, this means that they are forced to place orders infrequently, and the orders tend to be large.

This kind of buying pattern eventually created a "checkerboard" inventory in the store's warehouse. For example, suppose a complete bedroom set consists of a headboard, dresser, an armoire, night stands and a mirror. The store will have incomplete sets, which are missing, for example, the armoire and the night stand.

The result is that the consumer may skip buying the set altogether from that store, resulting in more lost sales. Another negative outcome is that the dealer is forced to make two deliveries to the consumer or is forced to take a special delivery from the manufacturer and pay excess charges.

Every time the retailer is missing stock, they miss some potential sales. By the end of a year, a great deal of retail sales potential has been lost. Orman Grubb pictures it this way:

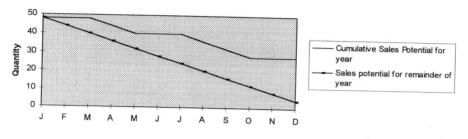

Figure 50   Lost Retail Sales Potential

As an example, assume that you have a piece of furniture that has an annual sales potential of 48 units (or, theoretically, 4 per month). You start out at the beginning of the year, stocked up with that item. In March, you are out of stock, and due to long lead times, you cannot get restocked until May. You miss potential sales of 8 units. Now, your annual potential is no longer 48 units, but 40 units. Each time you are out of stock, your annual potential drops.

The third customer cloud that Orman Grubb considered is a variation of the conflict between placing large, infrequent orders and small frequent orders for furniture. In this case, the reason for the large infrequent orders is to meet the dealer objective to keep freight costs down.

Dealers can easily measure freight charges as a percent of sales. Freight charges are "hard dollar" amounts. They are tracked by dealers and manipulated to make them as low as possible. What the dealers are not able to measure are the "soft dollars of margin" foregone due to lost sales. Because of this, the dealer concentrates on saving $500 in freight costs rather than gaining $2,000 in margin on sales he would otherwise have had.

The Theory of Constraints assumes that the goal of a business is to make money, not save money. The question that Orman Grubb had to wrestle with is how do you convince a dealer of this philosophy. Below, you can see the assumptions and injection that Orman Grubb decided on.

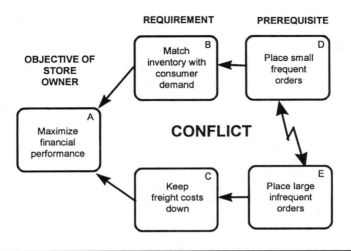

Figure 51   Cloud 3 — Buyer's Cloud of Freight Costs

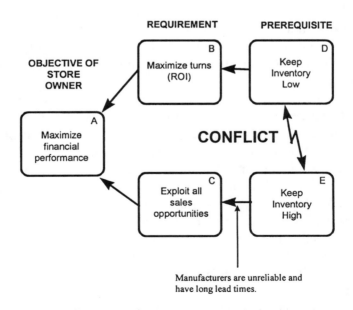

Figure 52   Cloud Injections — Cloud 1

Our assumption is, under the C to E arrow, that:

> *In order to exploit all sales opportunities, we must keep inventory high because manufacturers are unreliable and have long lead times (between the time the store orders goods and the time the goods are physically received complete).*

The injection that Orman Grubb implemented was to have a short lead time and a consistent arrival day. In the industry, lead times of between four to six weeks are normal. In fact, for some large east coast manufacturers, lead times of 10 to 15 weeks are not unusual.

Orman Grubb dropped the lead time to 2 weeks and 2 days, guaranteed. The two days provides some leeway for Murphy.

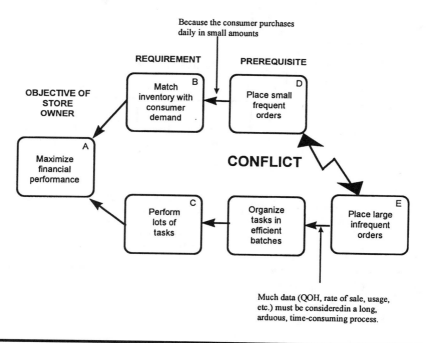

Figure 53   Cloud Injections — Cloud 2

The injection comes under the C to E arrow. The assumption is that:

> *In order to perform lots of tasks (i.e., complete the huge amount of work), the buyer must organize the tasks into efficient batches. In order to organize the tasks into efficient batches, the buyer must place large, infrequent orders because much data must be considered in a long, arduous, time-consuming process.*

Orman Grubb's injection is simple — make the process so easy that it takes up practically no time. The concept is to set a safety stock for the store on the Orman Grubb items. The order is always placed on the same day of the week, based on sales in the prior seven days. The stock is replenished on a 2.5-week cycle. No thinking is involved. All that the buyer must do is take actual sales during that seven day period and create a purchase order. No further thinking is required except a bi-monthly or quarterly reassessment of stock levels.

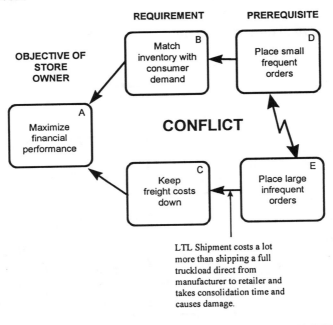

Figure 54   Cloud Injections — Cloud 3

The assumption under cloud 3 is:

> *In order to keep freight costs down, we must place large, infrequent orders because LTL shipment costs a lot more than shipping a full truckload direct from manufacturer to retailer and takes consolidation time and causes damage.*

The buyer is assuming that the furniture manufacturer is using a trucking company located close to the manufacturer that consolidates shipments from different companies and waits until they have a full truckload going to a specific location (e.g., from California to the east coast).

Or, perhaps even worse for the dealer's lost sales problem, the buyer waits to accumulate a "full-truck" worth of orders, further lengthening the time between consumer demand and product arrival.

The "Less-than-Truckload" (LTL) means of shipping is expensive and also incurs a much higher risk of damage. For example, the furniture would be loaded onto a truck at the Orman Grubb factory, and unloaded at the trucking company's warehouse. When the trucking company accumulates enough freight, the furniture would be loaded back onto another truck, and may be loaded and unloaded several times before reaching the retail store.

Furthermore, you could never be 100% sure when the consolidator would have a full truckload and be able to ship the goods. It could be this week, next week or the week after.

The injection is that Orman Grubb does the consolidation. They fill a truck where the first stop may be Arizona and the last stop may be Philadelphia. They have trucks going out every single day. This is much more cost competitive and reliable than the LTL method, yet does not require a full truckload quantity order to be placed by any single dealer.

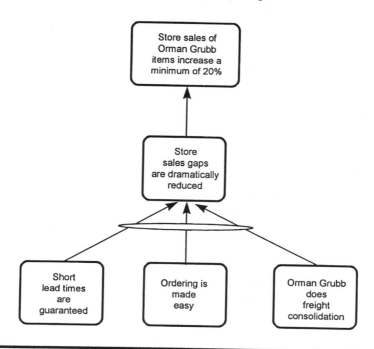

Figure 55 Results of the Combined Injections

If you combine the three injections together, Figure 55 shows the result:

As Jeff explains, "If you guarantee a maximum of two and a half to three weeks lead time from order to receipt of goods, and you make the ordering so simple, even a chimpanzee could do it. If we do the freight consolidation, then we can replenish stock so quickly that sales gaps are almost non-existent. The store is virtually never out of our stock!"

However, another astonishing element of the solution [which I believe was even a surprise to Jeff] relates to the frequency of delivery. Here is what happens to the tree:

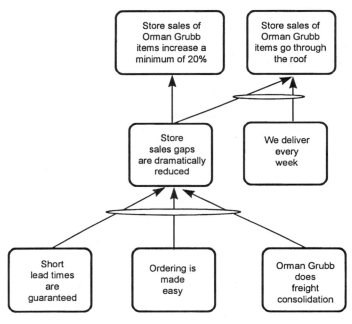

Figure 56   Frequency of Delivery

Look at the two new entities in the upper right of the diagram. The psychology of having an Orman Grubb truck at the door every week starts to impact the store buyers, sales people, and customers.

Orman Grubb has empirical data to back up this Future Reality Tree. As of March 1996, the company had worked in this model for nine months with three dealers and for six months with 11 dealers. They had 99 dealers on this plan. After a period of time on the plan, when a dealer is convinced of the reliability and consistency, the dealer sales of Orman Grubb items often shoot up as much as 100%.

Results

1. In less than a full year, Orman Grubb reports consistent dealer sales increases of between 20% and 100% for their items.
2. Orman Grubb credits this approach with winning them 37 new (additional) dealers in a two-month launch period. By explaining the store's problems to them and showing how a furniture manufacturer creates the problems, Orman Grubb's people have an entirely different relationship with prospective customers. They spend an hour with the customer discussing the new business method before they talk about product. They show the story of what their life is like now and what it's like after they implement the injections.
3. Price is much less of an issue with dealers. Traditionally, price was everything. As long as the price that Orman Grubb is offering doesn't insult their intelligence, it's OK, since the other benefits are so powerful.
4. Orman Grubb gets to pick and choose dealers. Their unique position in the market is allowing them to add profit by picking the dealers with much lower credit loss risk.
5. Collections have dramatically improved. Under the old paradigm, dealers would order large quantities infrequently. This meant that the dealer would get a huge invoice (e.g., $30,000) and not have that kind of cash in his bank account to pay the bill. With the dealers buying small quantities every week, the small bills are much easier to pay. The dealers pay more promptly because they don't want to stop the weekly delivery train. It's too good for their business.

Initially, the competitive reaction was one of denial. Competitors don't believe that anyone can do what Orman Grubb is doing, nor do they believe that it can have such an impact on the store's sales. Orman Grubb was initially concerned that competitors would react quickly, but more than a year after implementing the strategy, the competitors are still in denial.

The usual next phase for competitors is to try and copy a successful strategy. Of course, not knowing all of the pieces of the solution and not being TOC companies, Orman Grubb believes they will be less than successful.

Lead by a CEO who "barely graduated from high school" (Jeff's proclaimed self-description), Orman Grubb is growing dramatically. As Dr. Goldratt told Jeff at the conclusion of Orman Grubb's presentation to a conference in March, 1996, "I wish there were more people with an education like yours!"

Case Study 3: ABC Forge

This case study illustrates several very powerful lessons: (1) how vital it is to have senior management, including the CEO, directly and intimately involved in this kind of change process, (2) the ability to dramatically and quickly change the culture of an 85 year old company from the smokestack industry, and (3) the ability to gain significant competitive advantage by analyzing customer complaints to find the core problem — the disease causing most or all of these other problems.

ABC (not the company's real name) asked to have their identity protected for competitive reasons. However, we are very fortunate to be able to see their analysis of customer complaints (UDE's or Undesirable Effects) and the core problem they identified. This part of the case study is documented at the end of this chapter.

ABC Forge (ABC) is primarily a large metalworking and heavy capital equipment steel forge company supplying a variety of specialty forgings for the oil and gas industries. It is a privately held company with 300 employees, in business, since 1910.

In 1992, the company had an excellent quality reputation. ABC had invested heavily in technology and actually had a "working" MRP II system. Their job shop standard cost system was good, they had accurate labor reporting and had implemented a "Quality Improvement Process" using Juran techniques.

Unfortunately, all of these accomplishments were not sufficient to negate the effects of the market downturn in 1992. ABC was not making money, and the loss of their largest customer only added fuel to the fire. Layoffs made things worse.

In any environment of layoffs, any management effort to improve productivity and/or capacity is met with heavy resistance. Employees assume that the only result of such efforts will be more layoffs. Therefore, once layoffs start, it becomes more and more difficult to effect an improvement program. This was one of the key factors influencing the strategies of ABC.

There are three extremely valuable lessons to be learned from this case study. The first is that success requires strength in all of the security factors discussed in this book — sales and marketing, employee security, building customer value, etc. No matter how good you are at any one of these, if one of the necessary factors is weak, the whole organization doesn't produce. This, in fact, lies at the very essence of the Theory of Constraints. Second, it is not just the strength of the individual factors, but the way in which you align those factors and cause them to interact successfully together that makes the huge difference. Peter Cassidy, a senior manager at Nortel, describes this aspect of the Theory of Constraints as unique, allowing you to see the connections between functions, processes and events.

Third, factors such as an 82-year history in business and a strong P&L from last year are not good predictors of a company's future security and success. Rather, the factors lie more in the intellectual capital of the company, even in smokestack industries.

At ABC, several negative indicators were forcing them to start considering how to build their intellectual capital. Internal rejection rates were at 3.1%, deliveries had dropped to 40% on time and 20 to 35 day manufacturing lead times were being quoted with a 7 to 10 day response on quotes. The service reputation had deteriorated.

Simply put, the quoted lead times and on time delivery performance were noncompetitive. As a result, some of the smaller volume customers began looking elsewhere and sales declined.

Other undesirable effects included the inability to remain profitable during the market recession, the ongoing pressure to reduce operating expense (especially people expense) and the reluctance of stockholders to make future needed capital investments.

Around this time, the new sales manager gave the President of ABC a copy of *The Goal*. The principles were immediately put to work. The President and the VP & General Manager were trained in the Theory of Constraints and set about identifying their core problem(s).

The first core problem identified was in the production planning and scheduling system. It did not enable ABC to identify and manage the physical constraints in the plant. The most important role of the Theory of Constraints was to focus the problem solving efforts on a few key problems in a period of tight cash.

The Theory of Constraints production planning and scheduling methodology, called Drum, Buffer, Rope was implemented. Work in process was reduced and manufacturing lead times shortened. The plant expediter became the Drum Buffer Rope coordinator. At the same time, the Quality Improvement Process was used to increase Throughput out of the plant. The process was focused on the areas of constraint (e.g., rough-turn engine lathes, where capacity increased 175% just by this kind of focus and some subcontracting).

A significant change in culture was also adopted. The organization used to be a traditional hierarchy. In 1992, the decision to move to an open organization was implemented. This meant that every issue was open for discussion and any question could be raised without fear of reprisals. The objective was to create an environment whereby employee involvement is the rule and problem solving flourishes.

As the team worked its way through the problem, each internal constraint was identified and addressed. The constraint moved from the rough turn engine lathes to planer mills to forging presses to heat treat furnaces. Measurements were also identified and implemented to help management make the right sales and marketing decisions, as well as production decisions. Throughput per capacity constrained resource became an important measurement for quoting pricing, product profitability and order acceptance.

The impact that TOC had on the internally focused constraint improvement effort showed significant results. Here is what happened between 1992 and 1995:

	1992	1995
Internal rejections	3.1%	2.4%
On-time delivery	40%	75%
Inventory turns	6.5	8.5
Lead time	20–35 days	12–18 days

In fact, 93% of deliveries are less than a week late, beating the industry average of 88%.

Several measurements and survey results indicate the progress that ABC has made. One of the key measurements of productivity is T/OE — Throughput divided by operating expense:

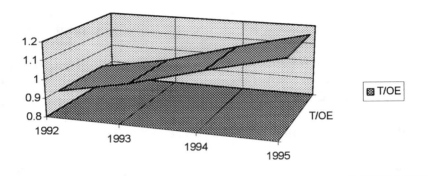

Figure 57   T/OE — Throughput Divided by Operating Expense

Another important measurement is Throughput per person:

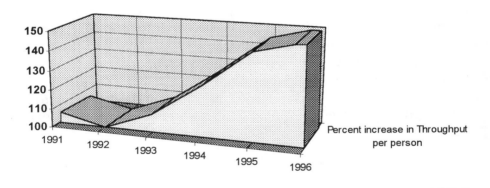

Figure 58   Throughput per Person

Another mandatory indicator that told ABC that they were on the right path came from customer-survey results. The survey asked customers to rank ABC compared to competitors on three factors — on-time delivery, quoted lead time, and general competitiveness. Following are the changes between 1992 and 1995:

	1992	1993	1995
On-time Delivery			
% above average	22	25	67
% average	58	65	29
% below average	20	11	4
Quoted Lead Time			
% above average	8	6	44
% average	62	68	43
% below average	30	26	13
Competitiveness			
% above average	20	24	62
% average	60	53	28
% below average	20	23	10

The key to any change process in a for-profit organization is ultimately to benefit the bottom line. The question was how to transform the new production responsiveness into sales. The TOC Thinking Processes were applied to ABC's current markets, and ABC used the Future Reality Tree shown in Figure 59 to initiate changes.

Note, in this first Future Reality Tree, how an employee profit sharing system, by itself, is not sufficient to cause employees to work towards company goals.

Note in the next page of the Future Reality Tree (Figure 60) the beginning of a market segmentation strategy. This tree documents the value of superior lead times, but only to some customers:

From here, ABC adds an injection that raises the Intellectual Capital as it is applied to Sales and Marketing. Specifically, they are looking for solutions to customer "core problems" rather than just symptoms, and to find better solutions than competitors (see Figure 61).

From here, ABC implements another injection (Figure 62) which emphasizes segmenting their markets, but at the same time using existing resources to more advantage. They look to choose market segments that are counter cyclical.

With this last injection implemented (Figure 63) more positive effects occur which signify that all of the original UDE's (undesirable effects) have been overcome.

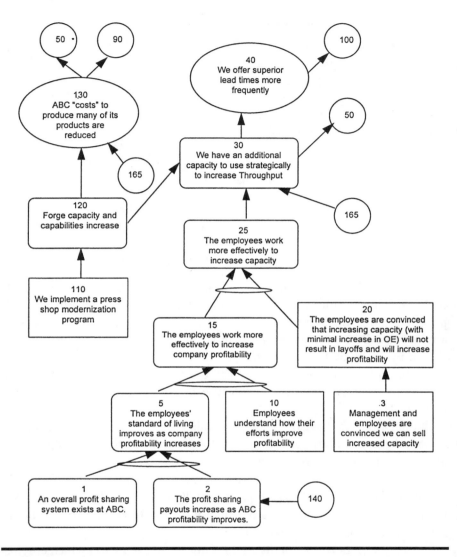

Figure 59   ABC Forge Future Reality Tree #1

In choosing the injections that they did, ABC was looking for two very key strategic effects. One was to create excess capacity downstream of their presses. The second was to reduce the risk of being dependent on one market. They no longer wanted to be market constrained when the market they deal with goes into a recession.

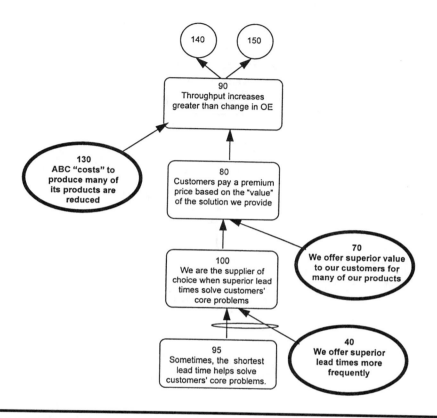

Figure 60   ABC Forge Future Reality Tree #2

Applying TOC to Marketing at ABC Forge

Similar to the Orman Grubb case study described earlier, ABC was cautious about collecting undesirable effects from customers. They undertook a process of nominating candidate customers. Then, they formed teams to collect customer UDE's. The teams consisted of inside sales people, the sales representative for that customer and a manager.

ABC then constructed a Current Reality Tree to identify what they felt were the customer's core problems. This was followed by construction of a Future Reality Tree where ABC strategically solves the problems. The Future Reality Tree was thoroughly scrutinized by the team.

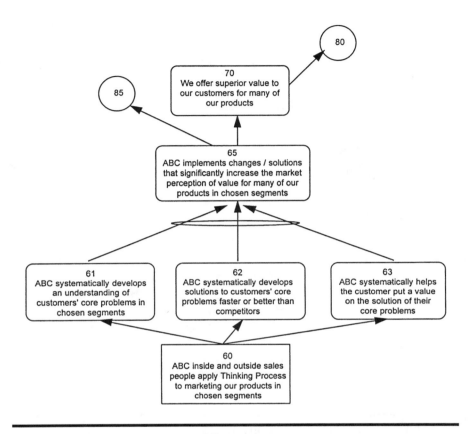

Figure 61   ABC Forge Future Reality Tree #3

The next stage in the process was the review of the Current Reality Tree with the customer, verifying the cause–effect relationships and seeking consensus from the customer on the validity of the tree.

After presenting the Current Reality Tree and the Future Reality Tree to customers, ABC would modify the Future Reality Tree to move closer to a full, practical solution to the customer's needs and problems. Each significant modification resulted in more presentations to the client, until ABC felt ready to draft an "offer" to the client. The idea of the offer was to provide a win for the client that would also hold long term benefits for ABC.

Once the Sales department had an offer in mind, they worked closely with other departments at ABC to identify obstacles and define intermediate objectives to complete the offering. This process is the Prerequisite Tree Thinking Process of the Theory of Constraints.

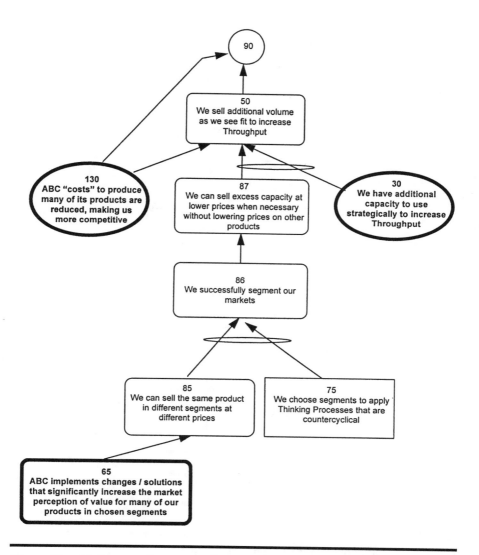

Figure 62   ABC Forge Future Reality Tree #4

The final step was to take the intermediate objectives from the Prerequisite Tree, identify the necessary actions required to implement the solution for the customer, and go and present the offer to the customer.

In selecting customer candidates, ABC looked at situations where there was significant potential to increase sales. They looked for customers whom they felt were very willing to think in a win-win manner.

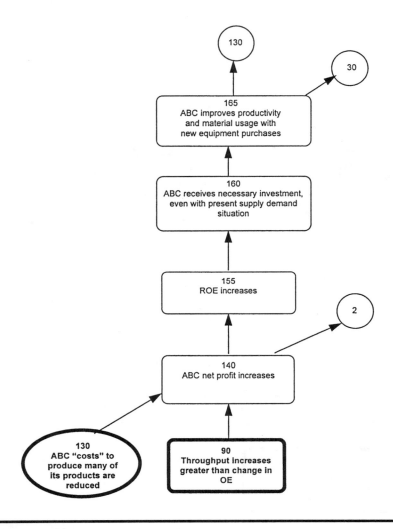

Figure 63   ABC Forge Future Reality Tree #5

Many of the changes that ABC implemented, as demonstrated above, required no capital investment. The results — significantly increased sales, protection of key customer profitability, etc. — were the outcome of changes in policies, procedures, training, professional development and internal measurements. Of course, all of these changes were orchestrated by people within the company who could see the bigger picture and could envision how the new ABC would work.

Changes such as Customer Emergency Response programs and estimation and pricing models don't require a lot of capital investment. Rather, they require you to understand how a few changes in your company can erase a whole range of negative effects in the customer environment.

Just understanding the importance of having a guaranteed, consistent and reasonable lead time for manufacturing can have an enormous benefit to customers.

ABC took the important step of educating their clients on the nature of their problems, and the contribution that ABC was making to solve them. Further, by presenting their work to selected customers, they are helping those customers improve their quality and constraint management procedures, thus enabling them to sell more and be more productive. This breeds long term customer loyalty while simultaneously addressing the constraint of the chain (your customer's production and quality constraints).

The Vice President and General Manager comments that "the President, the General Manager, and the Sales Manager, at a minimum, must be knowledgeable in the Theory of Constraints to be successful at driving the change process." For one thing, as can be seen from the injections described above, employees must know where the executive stands relative to downsizing and layoffs. While the Vice President agrees with the philosophy of "never downsizing", he believes that "never" is too strong.

"We have made a commitment not to lay off as a result of an improvement. Instead, we use these people to serve and support taking on additional market share," the Vice President explains. Of course, this required an employee openness to cross-training, so that ABC could better meet the needs of the market. This kind of effort, as described in the Scarborough Public Utilities case study above, is often resisted by unions and employees alike. Employees need to understand the benefits to them of undergoing training and potentially taking on more responsibility. They often view cross-training as something that removes the simplicity of their lives.

One interesting point that the Vice President makes refers to ABC's study of customer undesirable effects. As Dr. Goldratt believed, from a theoretical point of view, the Vice President actually found "many similar UDE's across the different industries we serve." By sincerely making the effort to collect these UDE's from customers and then sharing the Current Reality Tree, the Vice President confirms that these are "major steps in gaining customer confidence and buy-in."

In hindsight, the Vice President says that he "would have accelerated the TOC marketing and Buffer Management efforts," if he had it to do all over again. "DBR [The TOC plant scheduling solution — Drum, Buffer, Rope] is fast – in three to four months, our plant was full." Another hint that he offers is "don't shortcut the process. You will risk missing negative branches or uncovering obstacles that are in the way."

My interpretation of "not shortcutting," as the Vice President describes it above, is to make sure that you include as many intimately involved people in the process of scrutinizing the analyses. Also, it implies allowing enough time in sessions for people to raise their questions and reservations and answer them. I know, from direct experience, that this demands a great deal of patience, particularly from executives and managers who grew up in the world of giving orders and having them executed without question. To build and secure human capital for the long term, there is a significant educational investment in terms of involving these people in the analysis and decision-making processes.

Analyzing Customer Complaints for Competitive Advantage

In what follows, Figures 64 to 67, ABC company takes an actual customer (which we have disguised as "Hamco"), and uses the Current Reality Tree process of TOC to identify a core problem causing all of the customer's undesirable effects (UDE's).

If you look ahead to injection 60 and the result 61, on Figure 61 of the ABC Future Reality Tree, you will see where the following process is identified. The first task is to ask customers for their undesirable effects — the things that really bother them. From the information shared with ABC's sales and support staff, they must sift and find those problems that relate to ABC. Here is the list that ABC used from Hamco's input:

- Forging cost estimates are not always accurate.
- Product lead times are not always competitive.
- Some customers [of Hamco] perceive Hamco delivery promises to be unreliable.
- Not making an acceptable profit.

Using the resources at ABC who were knowledgeable in the Theory of Constraints, ABC sales and support people built a Current Realty Tree. The tree is very extensive. It is used not only for understanding how a few problems caused by ABC create a multitude of problems for the client. It is also an excellent presentation tool to show Hamco how deeply ABC has thought about their problem. Of course, ABC would not want to formally present such an analysis without having also thought about and come up with the solution.

As you will see from the analysis, every single problem that Hamco encounters stems from a few diseases such as "Forging cost estimates are not always accurate" (item 3 on the first page of the Current Reality Tree, Figure 64), and "Forging lead times are extending" (item 20 on the second page, Figure 65). From deeper analysis, these ares disease that ABC can fix, with some procedural and policy changes, and a new scheduling approach.

Here is the Current Reality Tree, showing how deeply Hamco is impacted by one problem.

All subsequent problems on this tree stem from item 3 — Forging cost estimates are not always accurate.

In this tree, item 20, Forging lead times are extending, gives rise to a plethora of undesirable effects.

In page 3 of the Current Reality Tree, Figure 66, entities 46 and 40, at the bottom of the tree, are both driving a lot of negative effects. Entity 46, that loss time is not made up, accounts for most of the undesirable effects.

As demonstrated in the results shown previously, the ABC team went beyond analyzing the problems. They came up with solutions that eliminated customer diseases and bred customer loyalty.

The Future Reality Tree provided the thread that wove all of the strategies together. This analysis of the customer problems would not have proceeded to a successful conclusion without the other injections — the resolution of capacity issues and lead time, the winning of employee commitment to increase capacity and productivity, and all of the other injections shown above.

Clients tell me that they receive visits from competitors and overseas diplomats, looking for the "secret" to the success of TOC companies. As you can plainly see, there are no secrets, just a lot of hard work.

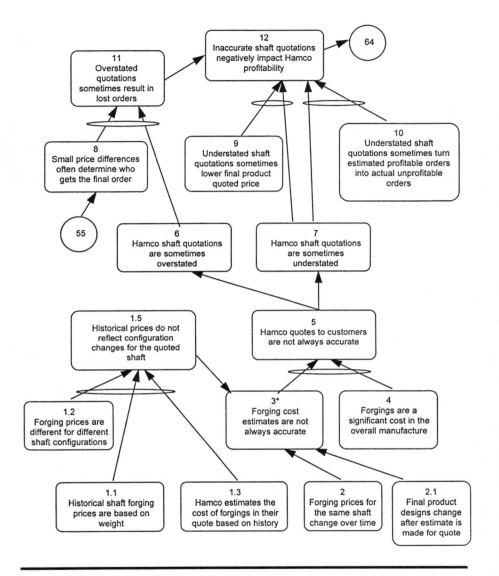

Figure 64   Hamco Current Reality Tree Analysis — Page 1

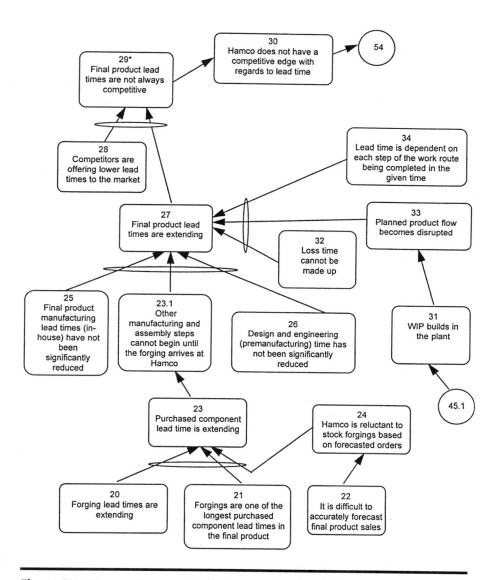

Figure 65   Hamco Current Reality Tree Analysis — Page 2

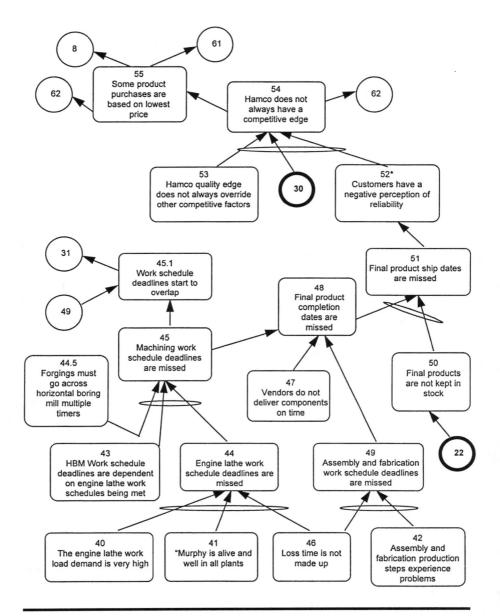

Figure 66   Hamco Current Reality Tree Analysis — Page 3

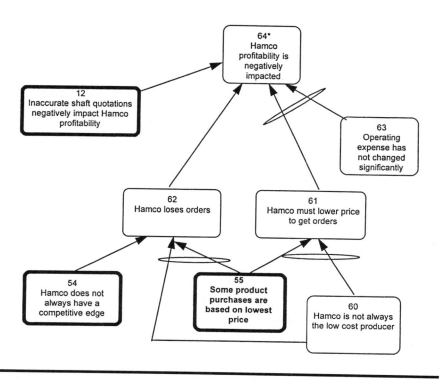

Figure 67   Hamco Current Reality Tree Analysis — Page 4

Case Study 4:
Strengthening Marketing Pillars at a Health Services Company

The health services industry is at war. The problem is, there are so many sides, the consumer can get hit from almost any angle. From my initial experience on the frontier of Canadian medicare in the 1960s to the present turmoil in the U.S. managed-care system, I've examined the problems from many angles.

It seems that I am meeting more and more doctors who are being driven out of their practices. Some feel that they just can't afford to pay malpractice insurance. Others feel the squeeze of managed care, and the pressure of insurance companies to cut payments to providers and prevent what doctors believe to be necessary procedures. The combination of these two factors and the results is scary. Some fine medical practitioners (idealists who entered the profession to help mankind) are leaving. If this trend continues, where will we be in 10 years?

The position that each individual portion of the industry has taken, in response to this turmoil, is to attempt to cut costs. As we'll see from this case study, better answers to the industry's problems are derived from a focus on Throughput, and on cutting the costs of the overall chain — not the individual pieces.

Several years ago, I had the privilege of applying some Theory of Constraints tools to a particular problem in one segment of the industry.

I have disguised some attributes of the analysis to protect the client. However, the lesson in this case study is the increased security, in a commodity market, of having several "pillars" of marketing to ensure ultimate success. Think of a "pillar" as one chain or one method of causing sales to occur. Think of a pillar as a support beam of a building. The more pillars, the stronger the building (to a point). You can never be 100% sure when one pillar will stop producing or when the competition might gain better control over one of the pillars. Therefore, more pillars ensures the building stays standing and also, if the logic is correct, causes more sales to occur.

As the nature of health services changes, it becomes clear that the key to long term security of a provider is how well they can collect and manage data and the ability to provide tools to turn that data into information which brings wealth and knowledge for people in the system.

As an example, in Canada, the government has put enormous pressure on service providers (physicians, laboratories, hospitals) to cut their costs. It actually reduced medicare payments for each service to providers, by huge percentages over a three-year period. The result was chaos, rebellion and a continued inability to gain control over the escalating health care costs.

In fact, there was huge potential to reduce costs simply by sharing information. If a patient is referred, for example, from a General Practitioner to a specialist, the specialist often orders the same laboratory tests all over again, at a huge cost to the government paid medicare system. If a patient sees several doctors for the same ailment, and receives multiple prescriptions for the same drug, the medicare system pays in full. There are many billions of dollars of documented abuse in the system. Yet by forcing individual doctors and individual laboratories to cut costs reaps a very small portion of the potential, and creates huge negative side effects in the process.

For example, ask yourself what is the logical outcome of the government reducing fees that they are willing to pay for lab tests by 25% in one year? The answer is — industry consolidation — mergers, acquisitions, and the like. The short-term result is unemployment from the downsizing that always follows a merger, and decreased competition which tends to lessen research into better ways of doing things.

Figure 68 is a simplified picture of the industry and the various links.

Through a variety of research projects, many funded through the entities seen in the diagram, new products, equipment, treatments and services are born. When these become authorized, having often worked their way through the myriad of government tests and regulations, they are commercially marketed through a variety of organizations. Drug companies have teams of salespeople who promote their products directly to physicians. Medical supply

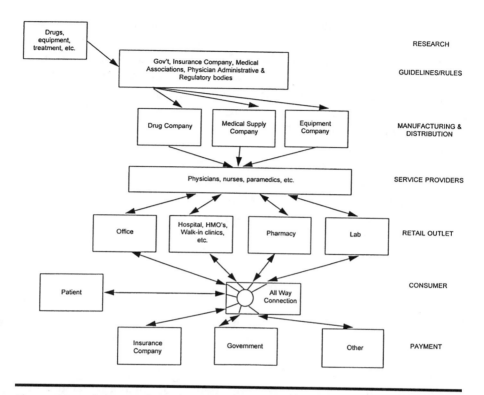

Figure 68　Diagram of a Health Services Company

companies may work more closely with large consumers of products, such as hospitals, HMO organizations, etc.

The service providers, such as physicians, supposedly make the decisions on what drugs to prescribe, what equipment to use, and what medical supplies will be consumed in providing service to a patient. In reality, insurance companies, hospital administrators, and government regulatory bodies are having more and more influence over the choices of the service providers.

The traditional way of providing services was either in a physician's office or patient's home, on a face to face basis. Increasingly, we see paramedics providing services at the patient's office (e.g., taking blood and urine samples and doing simple tests). In the near term, many such services may be provided by telephone, over the internet, or through some direct electronic hookup between a specialist and a patient who has hooked their body up, through self-installed electronic sensors, to an electronic network.

As to the question of who pays the bills and how the payment is consummated, this is becoming an increasingly complex issue. Ultimately, the consumer pays the shot. The question is, do they pay it through government taxes (such as for Medicare and Medicaid), or through insurance premiums or directly to service providers, cutting out the middlemen? In many countries that have tried a variety of options, we see a hybrid. Even in environments of total, state-provided medicare, such as Canada, the government eventually runs out of money due to abuse of the system. Then you see some services provided by state and some provided by the consumer.

No matter where you are in the chain, you can see that there are a wide variety of influences. The question of how to market to increase your sales and your long-term security is a difficult one.

My initial involvement was through an assignment with the Information Technology group of a lab services company. They were months behind in the planned release of new software to allow physicians to receive patient lab results online and integrate the results into a system that kept the history of patient tests. The Information Technology director wanted help on two fronts — one to determine what was holding back the new software project and second, to determine the components of an Information Technology marketing strategy for the company.

The first tool I used to understand the environment was the Current Reality Tree. After a few days of work, it became apparent that the new software would have no beneficial effects on the Throughput of the company. In fact, it would increase operating expenses for the development and support of the new system, while the old system also continued to be supported. There was absolutely no evidence that any of the thousands of doctors served would do any more business with this lab after the implementation of the new software package.

This became evident from discussions with doctors, researchers, and the sales people of this organization. It was also evident from the results of previous studies done by the four previous VP's of marketing, most of whom had not lasted one year in their jobs.

In this fiercely competitive marketplace, the market itself was not growing. By government mandate, the amount of money allocated to lab testing had been shrinking drastically over a period of a few years. The labs to which doctors chose to send their samples were based on factors that did not bear a lot of influence. For example, in many buildings where physicians practiced, a particular lab already had their offices set up, where they could collect the samples. For the convenience of patients, the physicians would rather send someone to have their samples taken in the same building. Another factor was lead time to get results. For the relatively few urgent situations, the physician would choose a lab whose schedule ensured that the results would be provided "stat" or urgently. In other situations where the analysis was complex, such as with the analysis of certain types of cultures related to cancer or other diseases requiring special analytical skills, the physician would choose the lab who had the best diagnostician.

In any case, the fact that a doctor would be able to get lab results into a more friendly software system meant absolutely nothing to their decision-making process as to which lab they would use.

Before looking at the Current Reality Tree of the physicians, I wanted to gain a better understanding of the Current Reality Tree blocking effective marketing at this company. I kept asking myself why the company went through several vice presidents of marketing in a few years. Figure 69 shows this Current Reality Tree analysis. The name of the company has been changed to XYZ.

In presenting this analysis to various management people on a national basis, I wanted their scrutiny on the validity of the logic. I facilitated a number of these meeting by teaching the group several rules of logic at the beginning of the meeting. Since the whole issue of how to spend marketing dollars was a very sensitive one, I was afraid that any meeting would become bogged down in everyone's opinions and emotions.

I therefore asked, at the beginning of the meeting, that the group use only those rules of logic that they had been taught to express their concerns about the analysis. If they had a question and couldn't ask it within the confines of the rules they were taught, they were allowed to address their comments to me as a process question.

This worked extremely well. I think that most people in each of the groups I presented to were relieved that they wouldn't have to waste time listening to opinions that didn't get them anywhere.

One of the first reservations I got was on the very first box, box 120, that claimed that XYZ did not have any kind of comprehensive process in place. Technically, the reservation is "Entity Existence," which means that "The box doesn't exist" or "That box is not true." Some people believed that there was a process, so I asked them to describe the process to me. After some humming and hawing, they quickly conceded that entity 120 does in fact exist.

By a process, I was referring to the idea that I could not identify a specific group of people, any objectives or any measurements associated with defining the strategy to attain more revenues through linking Information and

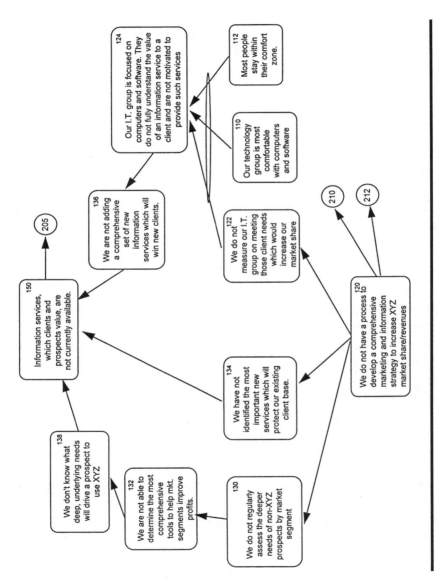

Figure 69 Current Reality Tree Analysis — Page 1

marketing. While the topic of how to increase revenues came up at various meetings around the company, there was no formal process to research what was blocking it and to accomplish it.

From this box, all of the negative effects stem. This first page dealt with why this effect leads to a lack of innovation from the Information Technology group, and to the lack of effort to understand various market segments.

The second page of the Current Reality Tree (Figure 70) deals with how these negative effects impact physicians, and how the field sales and support force was impacted and reacted.

Here we see how the natural inclination of Information Technology groups to focus on hardware and tools and the natural inclination of sales people to be suspicious of the technology combine to compound the negative impact on the market.

The final page in the Current Reality Tree (Figure 71) analysis shows how these negative effects continue to compound and combine with other facts of life. Finally, we see eight independent forces stifling profit and growth in this area of the company's business.

All of these effects and the logical cause–effect connections were verified, not just through scrutiny at management meetings but through interviews with field personnel, physicians and others.

Once I was satisfied, through feedback from a wide variety of people, that the logic was solid, I wanted to understand more about the physician's problems. I reasoned that any brilliant marketing strategy must ultimately address the physician's undesirable effects. This was not an obvious decision, since the end consumer was somewhat left out of the picture. However, from various interviews and readings, I concluded that the end consumer would not have the greatest influence on deciding which lab to use for the foreseeable future. This power still rested with physicians and others involved in the system.

In order to understand the physicians' points of view, I gathered information through a number of sources. I identified several physicians to personally interview. The company agreed to pay for the physician's time for the interview. I chose some physicians whom I had identified as leaders in their different views of how health services should evolve. I picked a variety of general practitioners, specialists, psychotherapists, and researchers. In addition, I attended a physician's focus group, sponsored by the company, where I prepared a set of survey questions. About two dozen physicians attended. Third, I received a computerized data base of responses to a survey by some 1,200 doctors scattered throughout various geographic regions. Fourthly, I

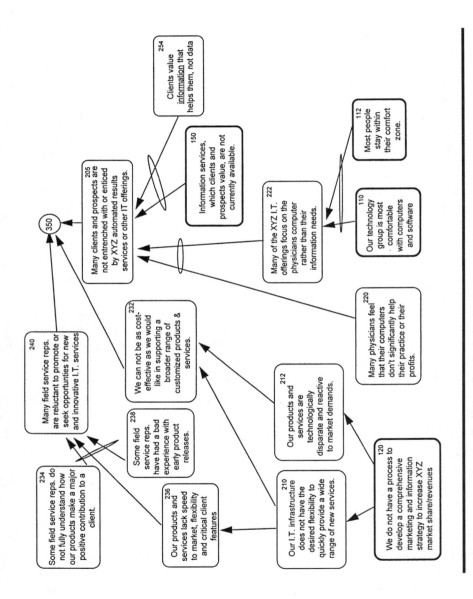

Figure 70 Current Reality Tree Analysis — Page 2

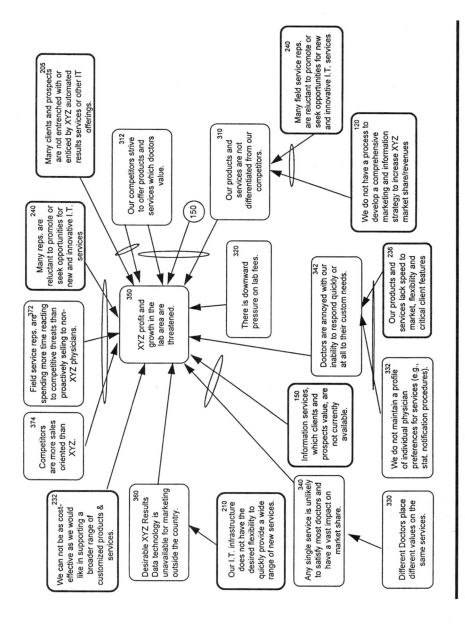

Figure 71   Current Reality Tree Analysis — Page 3

interfaced with some university researchers who were studying the medical services industry for some years.

From these sources, I formulated two physician's current reality trees and presented them to groups internal to the company. Both trees confirmed that XYZ company was not customer-focused on meeting the customer's information needs. Rather, they focused internally on their technology. This is what the Current Reality Trees of the Physicians looked like:

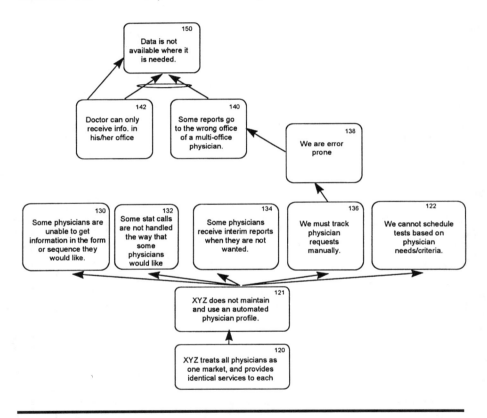

Figure 72   Physicians Current Reality Tree #1

While entity 121 was incorrectly supposing what a solution might look like, the essence of the problem is reflected in box 120 — a lack of individual customer focus. In fact, there was a driving force behind this. The Medical Director of the lab was always very concerned about the accuracy of the data provided to physicians. In addition to having to meet guidelines set by the medical associations, he was very concerned about legal implications if lab results were reported in different ways and different formats, and ended up being misinterpreted by physicians, resulting in harm to a patient. His answer was to have rigid standards where no modifications in format or style could be made to the information reported. Here is the second Physician's Current Reality Tree:

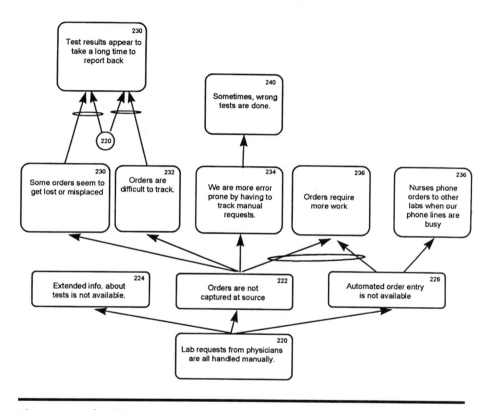

Figure 73   Physicians Current Reality Tree #2

With this analysis complete and verified, I wanted to understand why the company had gone through so many marketing VP's in a short time. Obviously, there were some conflicts that came up over and over again that prevented different marketing executives from getting the job done to the satisfaction of the CEO and senior executives.

I drew up a number of conflict diagrams that reflected on different parts of the Current Reality Trees, including the core problems. I then reviewed the clouds with one of the senior executives with a strong influence on marketing investment decisions. I wasn't happy with the response I received to any of the diagrams. Usually, when you have hit on the real conflict, you see an immediate strong reaction. In my case, I got some agreement on some aspects of the conflicts, but not the level of concurrence that made me comfortable. This VP did however encourage me by saying that the work and the diagrams were extremely interesting, and to keep him informed of my work.

Here is one of the conflicts that I presented:

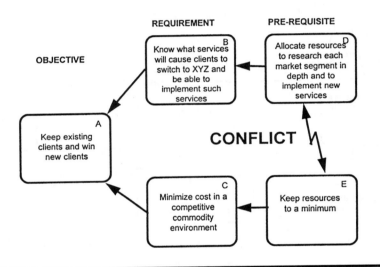

Figure 74   Comprehensive New Services Conflict Diagram (Assumptions follow)

Assumptions: B ← D

- Every market segment is different
- No one at XYZ has in depth knowledge of each segment

Assumptions: C ← E

- Marketing doesn't make much difference
- Previous marketing directors have failed
- Very few prospects will respon to our marketing efforts

Assumptions: D ← E

- There is no comprehensive plan/agreement on how to best allocate resources
- We don't have good data to allow us to predict the impact of new services on amrket share
- We can't predict that more marketing resources will help us

I pictured this conflict as occurring with each new Director of Marketing, who had ideas and wanted to spend money to pursue them. On the other hand, the executive was trying to figure out how to survive massive cuts in fees for services and still keep their share price at a reasonable level.

Another orientation of the conflict was as follows:

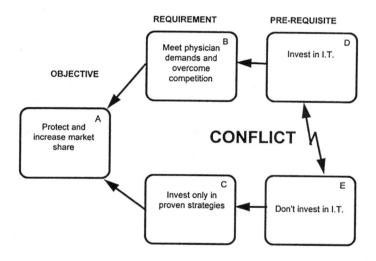

Figure 75   I.T. Investment Conflict (Assumptions follow)

Assumptions: A ← B

- The overall market is capped, and therefore keeping existing clients is essential to ultimately grow market share.
- Competition is fiercer than ever and will erode our client base if we do not keep our existing clients happy.

Assumptions: B ← D

- Some large XYZ clients frequently complain about competitive I.T. innovations which appear to be ahead of XYZ.
- I.T. services seems to be a way to differentiate ourselves.
- Some large client's are vocal and threatening with their demands for I.T. services

Assumptions: C ← E

- We have no proof that I.T. services increase market share.
- We have no explicit substantial proof that I.T. services are the major factor that protect market share.

INJECTION: C ← E   We identify and invest in new I.T. strategies which will convince, a priori, the XYZ senior management team of a high probability of driving up market share, e.g.,

Assumptions: D ← E

- There is pressure from clients and some pressure internally to invest in I.T., without proof from past experience that such investments are worthwhile.
- We view XYZ departmentally (e.g., Lab I.T., Real Estate, Marketing) and haven't agreed on whether to or how to use I.T. strategically to move the entire organization forward.

Assumptions: A ← C

- We do not like to take unnecessary risks.
- Keeping diagnosis data allows us to direct pharmaceutical companies to specific physicians for clinical trials. This causes more physicians to direct business to XYZ.
- Automated Lab software services are provided independent of which lab the physician is using. Physicians are willing to pay for these services because of their perception of value.

There were many assumptions that I reviewed with the senior executive. I could tell that some of them were painful to discuss. Also, since this was the only chance I had to meet with this one individual, it would have been difficult for him to be totally open with me.

From the various discussions and my increasing insight into this market and the company, I had some ideas about how a Future Reality Tree might look. I did some pilot work on some of the ideas, to verify their practicality, and then put together the Future Reality Tree. My approach was to come up with a lot of injections, to provide a high probability of success, were we to implement all injections. In fact, I knew that it would be totally impractical and completely unnecessary to implement all of them. My difficulty was not having enough intuition on which ones might win support within the company.

A total of 24 injections can be found on the following five pages of the XYZ Future Reality Tree (Figures 76 to 80). In combination, this was the first time that the organization had seen a recognition of the necessity to interact between the field, the Information Technology group, existing marketing personnel, outside third parties and the executive team in order to increase market share.

In constructing the solution, I relied heavily on the structure of the Current Reality Tree, identifying the desirable effects that needed to replace the negatives, and then asking myself what injections (ideas) were necessary to cause the desirable effects to occur. When stuck, I reverted back to conflict diagrams and discussed these with people to make sure my intuition was on the right track.

Here is how the Future Reality Tree evolved. The first page has eight injections to get the process initiated:

Most of these injections relate to assigning people and identifying the process by which they would do their work. The injection I100 at the top of the tree (Figure 76) suggests that widespread support for the strategies is required, and therefore a panel of "Key XYZ people" must be part of the final approval process. This is designed to eliminate the "local optima" syndrome of most Information Technology Groups.

The second page of the Future Reality Tree (Figure 77) suggests what ideas are necessary to implement the cultural changes within the Information Technology (I.T.) group.

One of the necessary effects that must be achieved is the freeing up of I.T. resources to work on strategic development. That effect, F226 (shown in the middle left of the tree) requires a special injection (I240) — a conscientious effort to examine all projects and decommit to non-strategic ones.

The next page of the Future Reality Tree (Figure 78) deals with market segmentation and field training. While these are two very different parts of the overall chain, both aspects must be addressed to have any chance whatsoever of a working strategy.

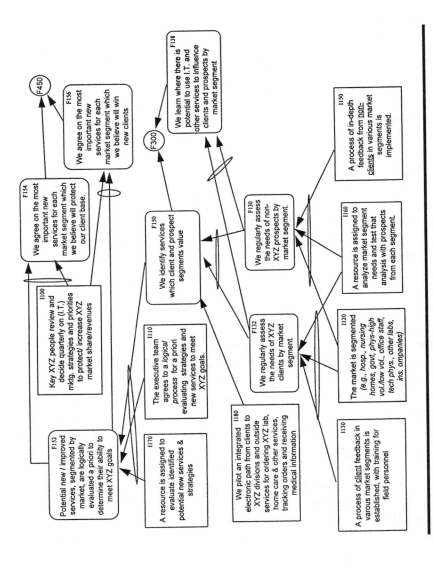

Figure 76 XYZ Strategic Marketing Future Reality Tree — Page 1

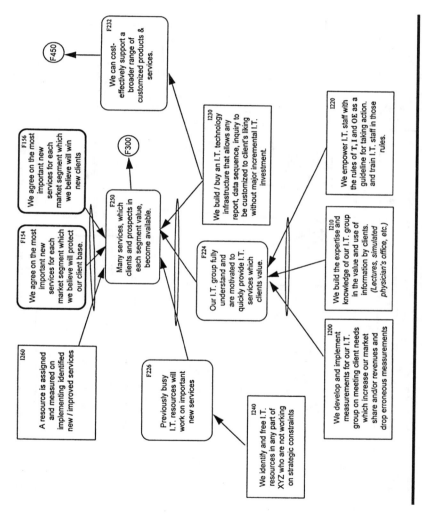

Figure 77 XYZ Strategic Marketing Future Reality Tree — Page 2

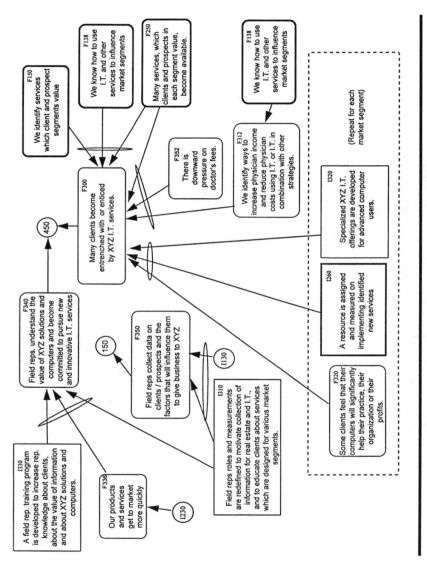

Figure 78 XYZ Strategic Marketing Future Reality Tree — Page 3

In this part of the solution, the dotted line around the bottom segment suggests that these injections must be repeated for each of the market segments that XYZ company wants to address. It may be the same resource in several cases. However, the specific attention of that resource to the different market segments is required. In addition, specialized offerings for each market segment are required.

Following on through the Future Reality Tree, you can see some of the elements of the Physician's Current Reality Tree addressed in this next page (Figure 79).

This part of the solution recognizes the value of customized information and solutions. Not all physicians are the same. I remember in some interviews, physicians were insistent that a lab report was useless to them unless it showed most current patient results first. Other doctors insisted that the opposite sequence was the best for them. The ability to easily customize to meet various doctors' needs is key.

The final page of the Future Reality Tree (Figure 80) is designed to exploit new technologies, such as the Internet, for information transfer and exchange. The value of relationships with third parties, such as various software vendors, is also recognized. Finally, the idea that executive measurements required a drastic change came to me from some discussions I had with certain divisions whose cooperation was in question.

From interviews with physicians, it became clear that any stand-alone information system containing information was useless to them. What they needed was a master patient record that contained all of the information about a patient and their medical history in one place. They told me that no matter how loyal they might be to one laboratory, they still consistently received results from different labs for a variety of reasons. Suppose a patient was in the hospital, for example, and had lab tests done there. Suppose a patient had their lab tests done at a lab close to work. The key, therefore, to managing electronic information was to have that information integrated into their practice management software.

Therefore, the injections shown at the top right of page 5 (Figure 80) of the Future Reality Tree take into account that there will be many different software solutions that physicians use. The key was to be integrated with a variety of them, and to exchange information with these software providers for a slot on their menus, or for some other useful advantage.

In the course of the study, I had received numerous suggestions for specific types of services appropriate to different market segments. Many employees

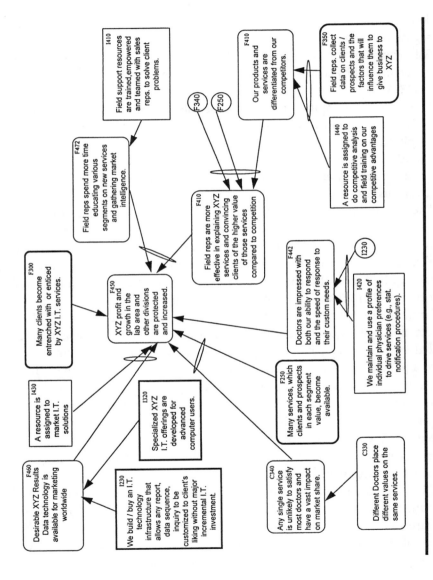

Figure 79 XYZ Strategic Marketing Future Reality Tree — Page 4

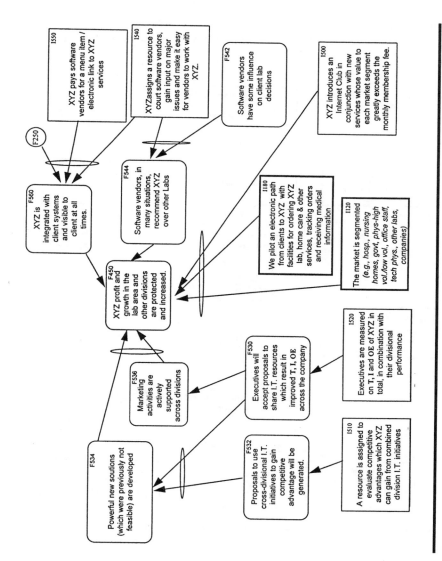

Figure 80 XYZ Strategic Marketing Future Reality Tree — Page 5

had thought a great deal about it, but had no outlet to make their suggestions come to reality.

Regardless of how valid this solution might be, I knew that the key to its implementation was how it would be communicated and sold to the various management layers within XYZ. I drafted the following transition tree for the I.T. director to present one of the injections to a regional management team:

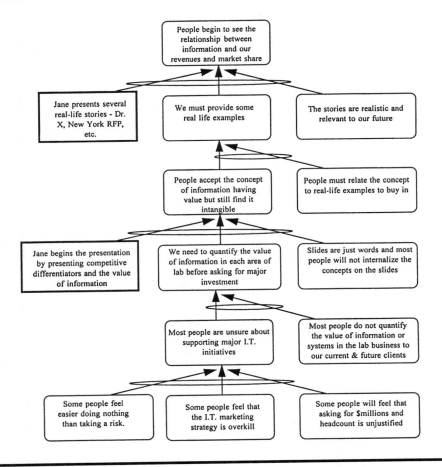

Figure 81   Getting Management Approval for an Injection

The meeting flow (Figures 81 and 82) is designed to recognize the widespread skepticism at the beginning of the meeting, and work to overcome it.

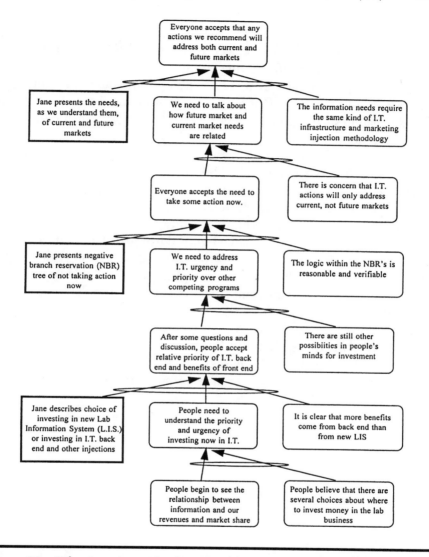

Figure 82 Why Investment in the Injection Provides Better ROI

In this part of the meeting, we must address the issue of choices for investment. In this specific instance, there were two capital intensive projects competing for investment. Due to the conservative nature of the senior manager, it was unlikely to get approval for both. Therefore, Jane needed to address the reasons why the investment in the injection would provide better ROI. In fact, the presentation of the Future Reality Tree, with the specifics of this injection, were the necessary material to convince the team.

If she could convince the team that her proposal was the better of the two, the next hurdle to overcome was the inertia of "let's not do anything." The problem with any profit-sharing incentive plan is the tendency to avoid taking risks, in the belief that if we don't spend any money, revenues will continue to roll in and we'll all collect a big fat bonus check. To overcome this, Jane had to show the negative side effects (Negative Branch Reservation or NBR) of doing nothing.

The final hurdle to overcome was why a major development, and not just a minor facelift, was required. The idea of flexibility to handle a variety of future needs and physician requirements is tied to having an infrastructure and information architecture, which takes some investment.

The last set of conditions necessary to gain approval relate to the cost/benefit analysis. Every proposal must be boiled down to T, I and OE. With this information in hand, and with good research to back up the probability of attaining at least as much Throughput as planned, Jane would be in a good position to ask for the management team's approval to go ahead with this injection (see Figure 83).

While this case study has been disguised to protect the company, it does closely parallel many situations that I've seen in my work with various industries. I hope that this case study will alert you to the need for actively building the intellectual capital necessary to have a secure organization.

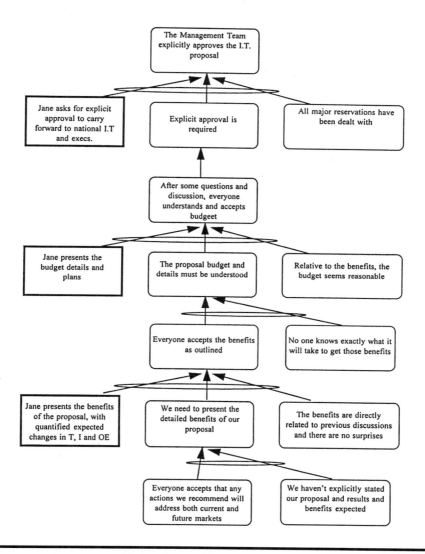

Figure 83   Why a Major Development Was Required

Case Study 5:
Applying TOC to Thousands of Employees at Alcan

Yvon D'Anjou is Vice President, Quebec Region, at Alcan Smelters & Chemicals. The company is a multi-billion dollar worldwide manufacturer of Aluminum products and byproducts. Based in Quebec, Canada, with plants located worldwide, Alcan has been very successful in capturing major world share of the aluminum market, and in keeping shareholders happy.

Yvon has worked in plants across North America, and has been a plant manager at several locations. He has become known within Alcan as a person who can get things done, particularly in the labor relations area. Yvon's most recent successes have involved extensive use of the Theory of Constraints.

The five thinking processes of the Theory of Constraints are designed to interrelate, each one being a "safety net" for the others. For example, if you miss something in the Current Reality Tree analysis, the Future Reality Tree will likely identify the missing pieces.

At the same time, there is no requirement to use all five Thinking Processes to analyze and solve a problem. For example, if you already have an idea that you are committed to implement, you can go right into Thinking Process 4 — the Prerequisite Tree — to figure out how to implement the injection.

This is exactly what Yvon did when he decided to help thousands of people in Alcan's factories in Quebec quit smoking. To understand the challenge, you have to picture a society where over 80% of workers smoke. Quebec, like certain parts of Europe, has not followed the North American trend to quit smoking.

With some 30,000 deaths each year, the cigarette is ranked as first among all causes of death and illness in Canada, according to the Canadian Medical Association. Among the victims, you find many non-smokers who suffer the effects of second-hand smoke.

When smoking is combined with certain toxic products found in the workplace, the combination can be even more harmful. Alcan believes strongly that a policy with related objectives on smoking is urgently needed. The ultimate goal is to eliminate altogether any employee exposure to tobacco smoke.

To make this vision a reality at Alcan, various departments and plants were asked to follow these steps (with some modification to reflect local conditions):

1. Help to stop smoking will be offered to those employees who want it and to members of their family.
2. From 1997 on, smoking will be prohibited inside Alcan buildings.
3. Each employee must receive complete information on smoking and its consequences.
4. Every new plant must be designated non-smoking.

The problem with this policy and the steps is that they do not provide a manager with a road map to go about it. Yvon decided to use the Theory of Constraints prerequisite tree approach, and discovered that there were 14 major obstacles to eliminating employee exposure to tobacco smoke.

With each obstacle identified, the corresponding condition (IO or intermediate objective) was identified that signified that the obstacle was overcome. Notice below the clever wording of the Intermediate Objectives.

There are some liberal translations in this document, which was originally written in French; e.g., intermediate objective 2 describes Smokers who have a *"bilan"* of their habits. In French, *bilan,* literally translated, stands for balance sheet. In this case, it means that smokers have taken an accounting or have a good understanding of the effects of their habits.

Once there was agreement on this set of obstacles and objectives, Yvon's challenge was to turn this list into a plan. In Theory of Constraints terms, this means creating a logical sequence in the form of the Prerequisite Tree. For example, there was no need to worry about employees becoming willing to help others quit (Obstacle/Intermediate Objective J21) before employees have all the pertinent information about tobacco related problems (A-1).

Elimination of Employees' Exposure to Tobacco Smoke

Obstacles	Intermediate Objectives
A. Employees hear a lot about tobacco-related problems, but do not know that much.	1. Employees have all the pertinent information about tobacco-related problems.
B. Smokers often do not realize all the effects this habit has on them.	2. Smokers have a "bilan" [accounting] of their habits.
C. Smokers see their habit as a fundamental right.	3. Smokers and employers know that there will be less and less space for smokers and smoking in the future. In conflict with quality of life.
D. Too often, non-smokers jeopardize their well-being to accomodate smokers.	4. The partners, employees, and representatives have empathy for smokers and want to help them.
E. Given the current trends, management alone cannot implement the change.	5. Employees and representatives, specialists and others become partners in the pursuit of policy achievement and target objectives.
F. Possible partners cannot help if they are not fully aware of tobacco related problems.	20. Possible partners know the complete situation about tobacco in the society.
G. To succeed towards a goal, the goal and objectives need to be stated. We do not have that.	13a. A draft version of the policy is available.
	17a. A draft version of target objectives is available.
H. The partners who are not involved will oppose the new policy.	13b. Partners have participated in finalizing the policy. The target objectives are stated.
I. Help for smokers to quit is not easy enough to get	14. The policy and target objectives are widely communicated.
	6. Some employees are trained to help smokers quit.
	23. Help is offered (re: quitting) to employees and family.
J. Employees are afraid to help smokers change their habits.	21. Some employees become willing to help others quit.
K. Non-smoking policy could be a very unpleasant surprise to smokers.	30. Informal hints are given to employees that, some time in the future, plant will become tobacco free.
L. Smokers need to see that overall smoking habits are changing.	31. More and more quit. There are less and less smokers at the facility.

Elimination of Employees' Exposure to Tobacco Smoke (continued)

Obstacles	Intermediate Objectives
M. Management and supervisors will see this endeavor as one unnecessary big monkey.	32. Management and supervisors receive information about the tobacco-related problems and are exposed to the empowered approach limiting their role.
N. During the transition, employees will wonder if the policy is in effect.	33. Progress is measured and communicated, showing the effectiveness of the policy and objectives.
O. Employees might receive negative reinforcement at home.	34. Families receive information. Family members participate in quitting sessions.

By creating the prerequisite tree, Yvon painted the road map that took a very complex, involved task and made it much easier to accomplish, like a paint-by-numbers approach. The Prerequisite Tree, shown below in Figure 84, is being used by 18 plants, with a total of 920 participating teams.

So far, several hundred workers have actually succeeded in quitting the habit permanently, with several hundred more undergoing the process. This affects not only those workers, but the employees around them, their families and friends. Therefore, without question, already well over 1,000 people have benefited from this wonderful program.

Yvon seems to feel ready for any challenge that will improve the lives of those around him. He has taken on challenges that have resulted in significant new business development and thousands of jobs throughout various regions in Quebec. Last fall, he helped the region recover from the disastrous floods of 1996.

Yvon is one of the few people I have met who captures the essence of the Theory of Constraints methodology. He takes responsibility for the results and blames no one for his current reality. He practices paradigm flexibility — he is ready to shift paradigms with no preconceived notions. For example, in our last discussion, he told me to think of union agreements in the future that could span much longer periods than traditional contracts. "5 years?" I asked him. "No, try 30 years!" Yvon replied.

With this kind of thinking, nothing will be impossible for Yvon.

ELIMINATION OF EMPLOYEES' EXPOSURE
TO TOBACCO SMOKE

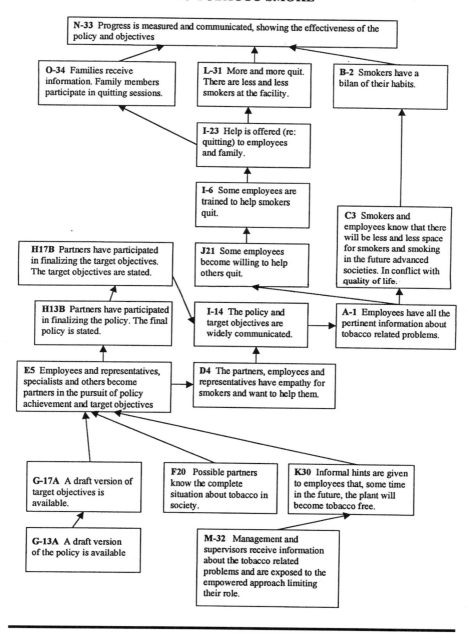

Figure 84 Elimination of Employees' Exposure to Tobacco Smoke

Case Study 6:
A High-Tech Company in Which the Operation was a Success But the Patient Died!

My very first Theory of Constraints study was a failure. The analysis was perfect, but the company executives wouldn't buy in. Two years later, they went bankrupt — an event that was easily predictable from understanding the injections that were mandatory for survival.

The lesson here isn't how smart I am. Rather, it's how we all look for shortcuts and sometimes, there are none.

I was consulting with a company I will call ABC corporation. They were a struggling high technology company owned by a European parent. They specialized in image processing — the storage and retrieval of documents and images from a system, with all of the elaborate ways of indexing and recalling documents.

Before I undertook my Theory of Constraints training, I was convinced that three ideas would make this company a winner in its field. They needed a new VP of R&D — the current team of two managers was a disaster. They needed a thrust of marketing programs — trade shows, advertising, etc. And they needed more salespeople in some of the bigger cities across North America.

By the time I applied my Theory of Constraints knowledge to their situation, I had found 18 injections (projects) that were absolutely vital for ABC's

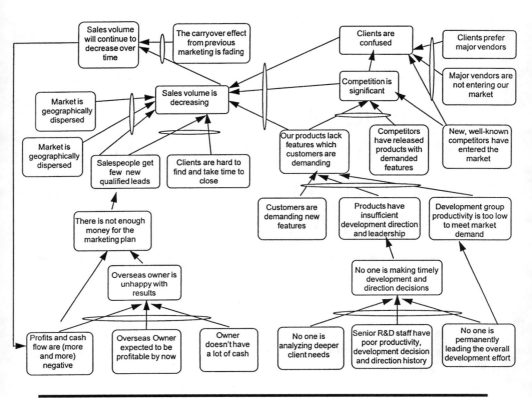

Figure 85 High-Tech Company Current Reality Tree

survival. While this made me a believer in the Theory of Constraints, I had no idea what kinds of challenges I was in for in terms of communicating my knowledge to the two key corporate executives — the President and the VP of Operations.

I presented the Current Reality Tree and got complete buy-in to my analysis of the problems (Figure 85). The Tree showed two main branches — one describing the interaction between the cash flow crisis and the lack of marketing, with its impact on sales. The other branch dealt with the Research and Development issues, which also impacted on sales.

Given the two branches, I analyzed two conflict diagrams — one on cash flow and one on the R&D side (Figure 86 and 87). Both of these conflicts had been going on for over a year, with no resolution in sight. I wanted to understand why.

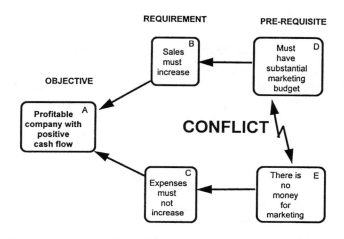

REQUIREMENT   PRE-REQUISITE

OBJECTIVE

B Sales must increase

D Must have substantial marketing budget

A Profitable company with positive cash flow

CONFLICT

C Expenses must not increase

E There is no money for marketing

Figure 86 Cash Flow Conflict Diagram (Assumptions follow)

Assumptions: B ← D

- We are covering a wide geographic area with 3 salespeople. Prospects are not numerous. It takes too long to find prospects without leads.
- Must generate good leads to make limited salespeople more productive.
- If Salespeople don't get good leads, they will not be able to earn sufficient income and will leave. This will become a vicious cycle of constant retraining and lack of productivity.
- Few prospects in the general audience know who ABC is or identifies with the name.
- ABC's product has more competitors than a year ago and it has not been significantly enhanced.

Assumptions: C ← E

- We must meet our payroll and other critical monthly bills
- It takes time to translate marketing into cash
- We can break even without marketing
- Marketing is risky

Assumptions: D ← E

- Owner is not sending additional cash

On this cloud (Figure 86), I worked on the D to E assumption. Underneath the stated assumption was the assumption that the owner had no confidence that any money spent would be returned. The owner had access to money through banks and investors in Europe and the U.S. The two executives running the company were both females in their first executive role, and were reluctant to ask for the money. They felt that it would make them look weak, because they had not been able to turn the operation around. The question I asked was, "Can you turn it around without any investment?".

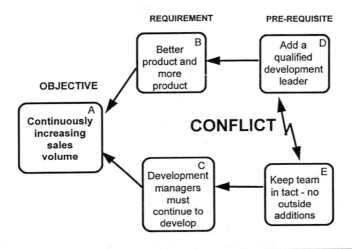

Figure 87   R&D Side Conflict Diagram (Assumptions follow)

Assumptions: A ← B

- Competition is fierce and forces better product.

Assumptions: B ← D

- There is no leader
- 2 developers are not producing or setting direction
- 2 developers do not have a good understanding of underlying client needs
- 2 developers will not necessarily leave company
- ABC will ultimately fail without excellent development leadership

Assumptions: D ← E

- Existing team may not have sufficient knowledge to carry on without current development managers
- ABC could fail if development team can't develop

- Development managers will leave company if we hire outside leader
- Development managers have helped clear up situations that other team members can't
- Existing team respects Development manager's knowledge — they may be upset if managers leave

This cloud, which seemed the most troubling because of the negative consequences of a bad decision, turned out to be easier to resolve than we thought. The two development managers were sent to Latvia to interact with the parent company. They met a senior American R&D leader over there whom they both respected a great deal. As part of a bigger development picture, he agreed to come over to North America and head up the development effort for 6 months, until the North American executive decided who would be the next VP of R&D.

Knowing this injection would be part of the solution, I still had a fear of some negative side effects, so I drew the negative branch seen in Figure 88.

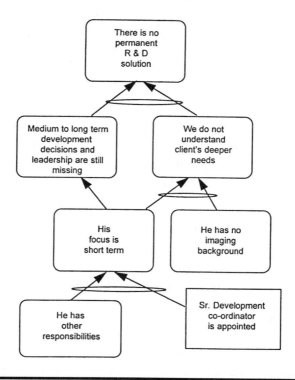

Figure 88   Negative Branch #1

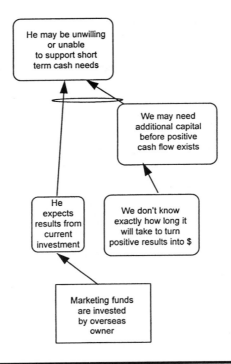

Figure 89 Negative Branch #2

I discussed this with the executive, and got their additional injections that they would define a permanent R&D solution and implement it in conjunction with the new Latvian coordinator's arrival.

The second negative branch regarding overseas investments proved more troublesome (Figure 89).

The first major resistance came from the President of the company. She liked the injection but refused to do anything about it. I couldn't understand why. She kept insisting that she wanted the company to prove itself, and get itself out of its problems without the cash investment. I took out the Current Reality Tree again, reviewed it, and started drafting a Future Reality Tree to prove to myself how relevant the cash issue was. After construction, my conclusion was that it was not only relevant, but vital to survival (Figure 90).

From this first part of the Future Reality Tree, it appeared that the cash issue was not significant. Since the parent company was loaning a high-level resource to ABC, and the company already had resources on board to implement

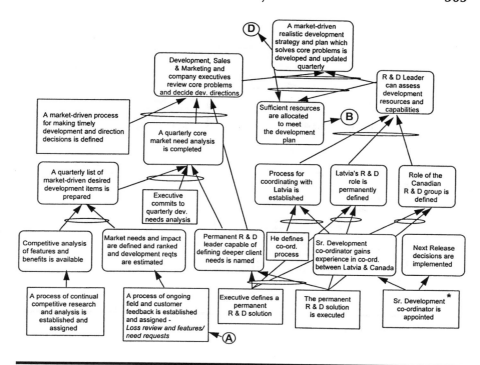

Figure 90   Future Reality Tree #1

the other injections, this part of the tree could be implemented without additional cash infusion.

The next part of the tree focused on the need to have some strategic relationships. Since the company did not have the direct contact with many customers who could use their product, the injections shown in Figure 91 focused on building relationships with other hardware and software companies that would be closer to the types of clients ABC corporation needed.

The last part of the Future Reality Tree (Figure 92) is what ultimately led to the company's demise and bankruptcy. Here is where the unmistakable need for cash is urgently shown.

There were other potential ways to get investment into the company, but not without the owner's knowledge and approval. For reasons which turned out to be very personal, the CEO simply could not and would not go overseas and ask for the investment. The end result was predictable — there was some progress on the R&D front, but without the needed boost in marketing dollars, sales continued to drop. The executives laid off employees until they couldn't lay off any more, and the doors were closed forever.

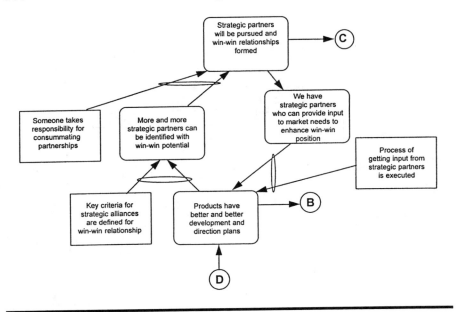

Figure 91   Future Reality Tree #2

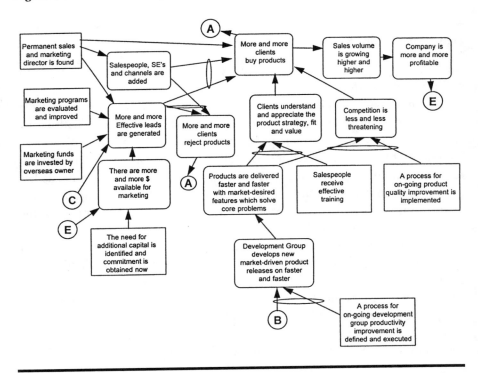

Figure 92   Future Reality Tree #3

The power of a Future Reality Tree is to help everyone concerned understand the minimum number of injections needed to cause the desirable effects to occur. After all, we have enough to do — do we really want to do more than the minimum necessary? Of course not!

However, if we don't do the minimum, it is not a matter of opinion as to the results. It doesn't take a brain surgeon to figure it out. For me, this was a brilliant success for the Theory of Constraints methodology, as far as analysis goes. It was a terrible failure at communications. I was warned, when I learned the methodology, that constructing a solution was half the battle. The other half was communicating it in a way to gain buy-in. I think this was wrong. I now believe communications is 80% of the battle.

Case Study 7: Acme Manufacturing Corporation

cme Manufacturing Corporation (not their real name) is a well-known consumer-products company in the process industry. They are a multi-billion dollar company, with healthy profits growing in the double digits. They are an international producer and marketer through manufacturing and distribution facilities across North and South America, and on several other continents.

This case study is included to illustrate several important aspects of Securing the Future:

1. How to deal with Performance Measurements
2. The importance of examining the entire chain and flow, from suppliers to end consumers.
3. The relevance of using multiple tools which are complementary to achieve order of magnitude improvement.
4. Some hints on implementing Throughput Accounting

I am grateful to two individuals for sharing their experience in the Acme Manufacturing efforts, which began with one division and is now embedded in several divisions. The names of the individuals have also been disguised to protect the company's competitive position. For the purposes of this case study, we'll refer to them as John and Andy.

John was a Business Unit Manager for Acme Manufacturing several years ago. During that time, he was charged with leading and working with a team to:

- Re-engineer manufacturing operations around Constraint Theory.
- Change organizational structure from a functional hierarchy to a team-based matrix structure.
- Change financial and operational measurements from traditional standards-based systems to Throughput-based systems.

Based on market, operational, and financial success with these changes, John was then asked to join Acme Manufacturing Corporate Manufacturing and Engineering to help implement these changes company-wide. John was part of this group, implementing these changes and involved in division turnarounds and manufacturing startups.

Andy, the Controller for one division of Acme Manufacturing, was a key player in introducing the Theory of Constraints to Acme Manufacturing. His experience played an integral role in eliminating the standard cost system and implementing Throughput accounting and the supporting business systems to cause the improvements.

Andy read *The Goal* (Dr. Goldratt's business thriller introducing the Theory of Constraints written as a novel) in 1989. Several months later, he heard Dr. Goldratt debate the two gurus of the Accounting field — Cooper and Kaplan. As CPA's and as well-known authorities to accounting professionals (through articles and books, including *Relevance Lost*) Cooper and Kaplan were instrumental in giving credence to what Goldratt had said throughout the 1980's — that standard cost accounting was "public enemy #1". In its place, Cooper and Kaplan were advocating Activity-Based Costing, which Goldratt was calling "the trillion dollar mistake". Dr. Goldratt was an advocate of something new, which he called Throughput Accounting.

Andy brought his boss to hear Dr. Goldratt speak at an AMA seminar in 1990. With what they heard, they started the drive to move from Standard Cost to Throughput Accounting at Acme Manufacturing.

John and Andy had several objectives to achieve for their business unit:

- Help the organization deal with and take advantage of synchronous flow
- Get away from Standard Cost
- Move from a hierarchical organization to a team based organization.

John told me that it took his team 2 to 3 years to get his unit turned around. He stresses that the team approach is critical in the evolution of operations, where this degree of radical change is involved. Though John led teams in the change process, he is adamant that it was *the team* that achieved the results.

He looked at the entire flow chain, from vendor to customer. In the process, he brought his business unit from a total supply chain lead time of 11 weeks to less than two weeks. While some of this was accomplished through changes from large lot to small lot ordering from vendors, the constraint management of the plant also drove a lot of positive results.

Other results included moving from nine weeks of inventory (raw material, WIP and finished goods) to two weeks.

The resulting customer lead time and reliability improvements boosted regional market share by 17 to 25%, depending on product line and market segment. An additional benefit of synchronizing all flow and speeding up inventory was the reduction in aged stock (over 90 days old) to less than one half percent of inventory. Net sales shot up 23%. There were many other positive effects, which are described below.

Buy-in to what John was doing came slowly at the corporate level. However, once the techniques were proven, the implementation went from one plant to company wide.

The change in approach and metrics was easily understood and embraced by operational personnel who "felt the pain" of traditional measurements. Upper management intuitively understands the intent and structure of the changes and so embrace them, but do not have to live with the mechanics and subsequent pain of the change process.

The hardest group to get to embrace the change is middle management. They bear the brunt of the effort for two basic reasons:

- They are the link between strategy and initiatives created by upper management that have to be implemented by first line operations. They make the "rubber meet the road."
- Many middle managers got to where they are in the organization by learning how to use and manipulate standards-based systems. These changes take their learned "tools" and throw them right out the window. This causes intense stress in people who are charged by their superiors to simultaneously run operations and completely change the way they "do business."

Andy explains, "Switching from standard cost to Throughput accounting is simple. There was no problem whatsoever in meeting GAAP requirements. You are dealing with the same numbers — just reporting them differently.

For example, you can take inventory and costs in aggregate. If inventory turns 12 times a year, you put one month's worth of expenses on the balance sheet."

"The tough part," Andy continues, "is the sales job. For example, the marketing people knew how to act with a standards-based system. They didn't know how to behave without standard cost."

"If you take standard cost out of the allocation to product margins, the immediate reaction you get from marketing people is 'Wow!! Look how profitable these products are! We can slash prices!' So you need to give people the whole picture. Standard costs don't go away. They show up in a different place in the P & L."

"Now," Andy concludes, "people can't even think in terms of a standard cost system."

John and Andy completed formal Theory of Constraints training in 1992. By that time, their change project was well underway. The idea of the TOC training was to help them implement better.

Why did Performance Measurements Need to Change?

Over the last fifteen years, Acme Manufacturing has gone through several initiatives intended to improve both profitability and customer service levels. From MRP II to ABC to looking at set-up reductions in the plants to challenging the validity of the standard cost systems, there was no shortage of effort to improve.

Unfortunately, most of these efforts were independently managed. The projects focused locally within departments. They did not provide the total business process integration solutions that the company felt were needed to be competitive and profitable in the future.

At the time, John and Andy visited several manufacturers to see what issues they were facing and how they were approaching change. Where appropriate, they also involved key managers where they thought the visit would help change "standard-cost paradigms" by seeing applied concepts in action. The only problem was that the people they visited only had a piece of the total solution. No one had implemented the total range of tools needed to effect change at a total organizational scale.

They found throughput accounting implemented without flow synchronization, throughput metrics overlaid on top of standards-based systems, flow synchronization in only pieces of internal and external supply chains and very few organizations who integrated the concepts vertically and horizontally through their total organization.

Virtually all of these manufacturers were being driven by these issues:

- Lack of understanding of and horizontal/vertical linkage to Acme Manufacturing goals, objectives and initiatives (no clear "line-of-sight" from operating levels). This consideration is identical to the one described above in the Scarborough Public Utilities Case Study.
- Results by accident, not by design (changes in performance results without changes in activity). People either got a "pat on the back" or a "kick in the butt" for month-end results, without feeling any "smarter."
- Product mix subsidization (effect of standard cost systems which allocate overhead to products, resulting in inappropriate product mix decisions and promotions).
- Standards variance analysis and use of fiduciary measurements/operating standards as sole tools to manage shop-floor performance. There is an illusion here of precision, when in fact there is no focus or ability to use these reports to improve overall throughput. For example, scrap and loss rates were built in. The focus was on variances to frozen standards rather than on cost drivers.
- Planning assumptions based on infinite capacity

The symptoms of these issues included:

- Local vs. global measurements, resulting in sub-optimization, fire-fighting and short-term performance focus.
- Non value-added activities and associated costs as existing measurements focused scarce resources on standards maintenance and variance analysis rather than on true organizational effectiveness issues and drivers.
- Compliance organization, lack of initiative, complacency, status quo and lack of commitment at the operating level.
- Survival rather than continuous improvement.
- Volume and cost focus rather than quality and service.
- Sophistication and computerization to "control" chaos rather than root cause analysis and resolution.
- Measure everything (quantity of measurements rather than quality to ensure efficiency).

From this fragmentation, Acme Manufacturing recognized the need to approach measurements from a total organizational perspective. They quickly realized that there was no single "magic-wand" solution to their problems available on the market.

The Measurements and Vision That Acme Manufacturing Changed To

When you look at the before and after picture, it is apparent that Acme Manufacturing went through a major paradigm shift. For example, here is part of the before and after vision that was presented to employees:

Traditional	Conceptual
Sell capacity	Sell capability
Sell what we make and limit SKU variety	Make or purchase what consumers want to buy
Distribute only our product	Distribute any product going through similar channels
Multiple manufacturing divisions aligned by product and/or manufacturing technology	Division(s) aligned by market segment
Large customers with similar needs	Large customers with dissimilar needs
Generic products	Combination of generic and tailored products
Long production runs	Production runs matched to demand
Manufacturing value-added close to supplier	Manufacturing value-added close to customer
Product innovation	Process innovation
Divisional/group sub-optimization of capacity and capability	Total company optimization of capacity and capability
Individual/job-oriented organization structure	Task/function oriented organization structure
Standard costing/gross margin focus	Throughput accounting/incremental revenue
Tactical internal focus	Strategic external focus
Division and product focus	Market and capability focus

This vision was backed up by a mission that stated, "The mission of the Leadership Team is to create and promote strategic goals and initiatives." This includes:

- Defect-free products
- Injury-free work places
- Fast and consistent service
- Rapid commercialization of new products and processes
- Cost leadership
- Customer focused
- Team work — a way of life

From this vision and mission, Acme Manufacturing developed the following criteria. Every measurement must:

- Support **strategic** goals and initiatives.
- Be simple, relevant, understood and **actionable**.
- Add value from a **customer's** perspective.
- Positively affect **profitability** in the short and long term.
- Drive **collaborative** effort.
- Drive **continuous improvement** of market, operational and financial performance.

Acme Manufacturing did not find the ultimate solution that you can just drop into any location. The generic measurements must be tailored to the specific market, operational and financial needs of each division to ensure relevance. As a result, the measurements listed below are representative of the scope (depth and breadth) of measurements implemented in various divisions worldwide. Core performance measurements, however, are common and are rolled up to plant, division, group and corporate levels as appropriate.

The measurements allow Acme Manufacturing management to view all functions from an integrated and holistic perspective by providing a means to balance inter-functional customer service, financial and operational priorities.

In using the measurements, trends and relative rates of change are more important than absolutes. Examples include:

Customer Service (competitive advantage)

- Market performance by segment (share, growth, threats/opportunities)
- Percent of orders/line items delivered on time
- Percent of orders delivered as requested
- Customer "order-to-dock" lead-time reduction percent
- Number of customer complaints by category
- Percent of product shipped commercialized in last X years
- Percent of product shipped under customer special service programs
- Number of times we said "NO" to customers (and reasons)

Financial Productivity (creation of wealth)

- Throughput (Sales dollars minus raw material)
- Throughput per constraint hour (mainly as a mix evaluation tool)
- Inventory (valued at raw material content)
- Operating expense (actual non-inventory expenditures)
- Net Profit (Throughput minus operating expenses)
- Return on investment (Net profit divided by total assets)
- Productivity (Throughput divided by operating expenses)
- Inventory Turns (Raw material content of sales divided by total inventory)
- ABC item cost drivers (80/20 analysis)

Operational Productivity
(manpower, material, machines, methods, and money)

- Operating equipment effectiveness (OEE/UEE) capacity =

Utilization	×	Efficiency	×	Yield	=
Time running		Average run speed		Sellable output	
Total time available		Design speed		Total output	

■ breakdowns	■ policies/procedures	■ process quality
■ lunch/breaks	■ material quality	■ material quality
■ setup/changeover	■ process quality	■ machine capability
■ weekends/holidays	■ maintenance	■ test samples
■ stockouts	■ work habits	■ machine maintenance
■ product mix	■ paradigms	■ work habits
■ scheduling	■ learning curve	■ learning curve
■ maintenance	■ product mix	■ setup/changeover

- P-time (customer order, manufacturing, supply and new product development)
- inventory integrity, availability, age and days on hand
- safety, housekeeping and environmental statistics
- capital improvement and projects to remove bottlenecks
- scrap, rework and repair
- setup/changeover
- controllable expenses (overtime, supplies, insurance, travel, etc.)

The more significant measurement changes by function resulting from the above include:

- **Sales and Marketing** — "Throughput $" and "Throughput $ per constraint hour" as performance measures rather than the traditional volume and standard cost contribution margin.
- **Manufacturing & Distribution** — Quantitative OEE/UEE-based capacity management/project prioritization, inventory value/time period coverage/age/integrity/availability management, controllable operating expense driver management and P-time/supply chain management/flow synchronization.
- **Finance & Accounting** — Financial driver management as a resource rather than standards maintenance/variance analysis as a marketing and manufacturing referee.

How the Changes Were Implemented

John said that Acme Manufacturing started with one of the more progressive and profitable divisions. The change process took one and half years to implement. All of these changes were handled with internal resources. As they gained experience in using the measurements in the pilot division, and dramatic positive results were quantified as a direct result of the new measurements, they were better able to expand the concept to other divisions.

A key element in the success has been communication and training of concepts and measurements. Until understood and actually used, measurements are very intangible to most people, and also very difficult to sell. A lot of time was spent in meetings, training classes and one-on-ones explaining, applying and comparing the new measurements to the old measurements.

This is something that John would not change. He believes that this time is well spent.

Financially, this included the development of a throughput income statement and sample product mix change/operational exercises to illustrate the difference. Operationally, it included detailed standards vs. OEE capacity analysis and capital plan/ROTC implication comparisons to understand.

Through the education and awareness process, key division, group and corporate managers reached a "critical-mass" knowledge level and were willing to make a leap of faith to embrace the new measurements. This became easier as the implementers were able to show in-house quantitative improvements as a result of the measurement changes.

Initially, John and Andy recommended running existing and new measurement systems parallel (especially financial systems), to provide a fall-back and to ease apprehension. In retrospect and in current practice, they do not recommend running parallel systems for two reasons:

1. The concept is now proven and there are enough resources to help divisions as required.
2. There is a tendency in people and in functions to ignore the new system in favor of the old system until the old system's plug is about to be pulled. This negates any learning opportunity that may have existed.

Benefits Experienced

Acme Manufacturing has seen dramatic profitability and customer service improvements as they better focus on true customer service, operational and financial drivers, as evidenced by dramatic improvements in division, group, and corporate income statements, balance sheets, stock price and cash flow.

Here are some selected worldwide division improvements, after 12 to 18 months into the change process:

Market share	17–25%
Net sales	23%
Customer complaints	47%
Customer order leadtime	80%
Total stacked lead time	50–90%
Fill rate (promises kept)	88%
New product % of sales	50%

Manufacturing capacity	50–300%
Total-process steps	50–90%
Inventory days on hand	50–75%
Scrap reduction	32%
Housekeeping	20–40$
OSHA recordables	86%
Capital deferment	$MM through capacity increases

The operational benefits have been significant, making it easier to do business both internally and from a customer's perspective. These benefits cannot always be expressed in strict financial or statistical terms. For example, there has been a reduction in conflicts between departments and operating units. Everyone is marching to a common drummer and the conflicting measures have been highlighted and eliminated. Also, with these measurements being easy to understand and communicate, people commit to them rather than just comply with them.

Now, there is a profit mentality at all levels and across all functions, with a line of sight to customer service goals. People are constantly striving to identify and focus on all constraints — external or internal — that are blocking improved performance.

Accounting practices have been grossly simplified, eliminating many non-value-added functions. People are more effectively deployed, which reduces costs.

Acme's results are proof that these strategies are continuing to pay off. The fact that the change revolution was completed with internal resources speaks to the idea that most organizations have the right answers. They just need some tools to unleash their intuition and come up with these answers. This is where the Theory of Constraints and its complementary nature to other tools holds a great deal of value.

We don't need smarter people. We just need to teach people how to think.

Glossary

Assumption — A statement, belief or condition by which we model our world. We accept our assumptions as facts. Recognizing which of our assumptions block improvement is a key role of the Theory of Constraints.

Categories of Legitimate Reservation — Eight checks applied to verify that a diagram, a statement or a series of statements "makes sense." The checks are scientific-like principles of logic. In layman's terms, they are:

1. **Clarity** — Is every statement clear, and are the connections between statements clear? For example, the statement "Computers are a pain" can mean a lot of different things. The statement "Upgrading to new computer software is very time-consuming" is much clearer.

2. **Entity Existence** — Is the statement true? For example, the statement above may be very clear, but may not be true. In fact, we know that some upgrades are easy and others are very difficult. If someone raised an entity existence reservation, we might restate the entity as "Some upgrades to new computer software are very time consuming."

3. **Causality** — Does one thing really result from the other? For example, "If you smoke cigarettes, you will look cool". While the tobacco companies would like us to believe this connection, it is subject to a causality reservation.

4. **Insufficiency** — Does that statement, all by itself, cause the next result to happen, or do two or three things need to all happen together to cause the result? For example, "If I turn off the light, then my room is dark". In my room, I would have an insufficiency reservation. The correct statement is "If I turn off the light and I close the shades, then my room is dark".

5. **Additional Cause** — Are there other, independent and important causes of the result? For example, in the above situation, an independent cause of my room being dark is if it is nighttime and the power went off.

6. **Predicted Effect** — A technique to validate or invalidate the logic, by showing how the existence of something else proves or disproves the logic or the statement. For example, suppose someone makes the statement "O.J. is struggling financially". I could use the predicted effect reservation to challenge the statement by saying, "If O.J. were struggling financially, one could predict that he would not be looking at buying multi-million dollar mansions in Florida."

7. **Cause-Effect Reversal** — What causes what? For example, "If he needs more knee surgery, then his doctor screws up." Did the need for more surgery cause his doctor to screw up? No. The correct cause effect logic is, "If his doctor screws up, then he needs more knee surgery."

8. **Tautology** — Does one entity really cause the other one, or simply explain its existence? For example, "If there is are ambulances on the highway, then there is a car accident." The ambulances did not cause the accident to occur. They simply explain how we know that there has been an accident.

Chain — The entire set of interdependent events and interrelationships that must occur or exist to achieve a result.

Cloud — A five box diagram that defines the elements of a conflict. It is also referred to as a conflict resolution diagram. It is a structure of necessary condition, meaning that the elements to the right of an arrow are necessary to lead to the entity to the left of the arrow, but may not be sufficient to cause the result.

Constraint — Anything blocking us from moving closer to our goal.

Core Problem — A single, underlying cause of the majority of undesirable effects in a chain or in a Current Reality Tree. The disease causing the majority of symptoms.

Current Reality Tree (CRT) — A diagram that provides the focus of our improvement efforts. Through cause–effect logic, it shows the connections between the undesirable effects (symptoms) in a given environment and the core problem leading to those undesirable effects. The CRT answers the question of "What to Change?"

Desirable Effect (DE) — A positive or beneficial outcome. The "opposite" of an undesirable effect. Desirable effects appear in Future Reality Tree diagrams. They may be exact opposites of their corresponding undesirable effects or they may be conditions that signify that we have overcome the undesirable effect. For example, the undesirable effect may be "My friend is dying". While we would love to have a desirable effect that "My friend is not dying", that may not be possible. Our desirable effect might be "I can cope with my friend's illness".

Entity — Statement contained in any given box of a logic tree.

Evaporating Cloud — The thinking process used to diagram the conflict which is blocking us from overcoming a core problem. It shows why the core problem was never solved and fosters a new, breakthrough idea. This idea is the beginning of the answer to the second question of change — What to change to?

Future Reality Tree (FRT) — The strategic solution to our core problem, identifying the minimum projects and ideas necessary to cause improvement. The FRT is a cause effect diagram that allows us to test the validity of actions and solutions before proceeding with a proposed idea making sure the idea will eliminate undesirable effects without creating any negative side effects. It answers the question of "What to Change to?"

Goal — The place you run to with the football. The goal can represent any objective that you wish to achieve. The Theory of Constraints is not limited to achieving monetary goals.

I — See *Inventory/Investment.*

Injection — An idea or project that must be implemented in order to achieve a desired effect (as in a FRT). A breakthrough idea in a conflict diagram, that will remove the conflict permanently.

Inventory/Investment — As defined in the Theory of Constraints, it represents all of the inventory in the system (from raw materials through finished goods) and all of the capital investment used to create Throughput.

Invisible Constraint — Something that is blocking you from improving that you cannot see, feel or smell. It typically originates from a policy or inappropriate rules, training or measurements. Invisible constraints are very difficult to identify.

Negative Branch — An undesirable development that results from a proposed idea or action. When encountered from injections in an FRT, another injection(s) must be added to avoid or overcome the feared negative.

OE — See *Operating Expense.*

Operating Expense — All of the expenses which are used to turn Inventory into Throughput. This includes labor, sales and marketing, advertising, operations, depreciation, etc. Typically, the only expenses not included are the direct, per unit cost of materials in a product or service sold.

Physical Constraint — Something that we can see, feel, smell that is blocking us from improvement. For example, a machine in a plant that can not process parts as quickly as the market demands, or not enough people to complete a process on time. Physical constraints are often easy to spot because of the backlog of work in front of them.

Prerequisite Tree (PRT) — The detailed plan of all the obstacles we need to overcome to implement the ideas and projects in our Future Reality Tree. Obstacles are translated into milestones (intermediate objectives) which are then sequenced in diagram form. The diagram shows which milestones must accomplished as a prerequisite to accomplishing the others. This diagram, in combination with the Transition Tree, answers the third question of change — How to implement?

Root Problem — An entry point in a current reality tree that leads to any undesirable effects.

Rules of Logic — See *Categories of Legitimate Reservations.*

T — See *Throughput*

Theory of Constraints — A methodology to take any situation and improve it.

Thinking Processes — Five diagramming techniques using logic to answer the three questions of change — What to Change, What to Change to, and How to Implement the Change. The five processes are the Current Reality Tree, Evaporating Cloud, Future Reality Tree, Prerequisite Tree, and the Transition Tree.

Throughput — Total revenues minus directly variable cost of sales (usually raw materials). It represents the dollars that we bring in from every unit of sale, after paying the direct suppliers of the raw materials and purchased parts.

Transition Tree (TRT) — The detailed action plan, and corresponding logic, that we need to undertake to fulfill our plan. It might describe the necessary actions leading to an intermediate objective in the Prerequisite Tree or those actions required to implement an idea in our Future Reality Tree. This diagram, in combination with the Prerequisite Tree, answers the third question of change — How to implement?

Index

N

O

P